American Philosophical Society Library
Publication Number 11

Natural History and the New World, 1524–1770: An Annotated Bibliography of Printed Materials in the Library of the American Philosophical Society

ANITA GUERRINI

American Philosophical Society Library · Philadelphia · 1986

Cover illustration: "Corne of Asia," from John Gerard, *The Herball or Generall Historie of Plantes* (London, 1633). Deposited in the American Philosophical Society Library by the Pennsylvania Horticultural Society.

LC: 86-071010
ISBN: 0–87169–380–1

For M. A. O.

Contents

Foreword

The Andrew W. Mellon Foundation, seeking to increase the scholarly usefulness of our collections and enrich the intellectual life of the Library, made a grant in 1982 to the American Philosophical Society establishing a fellowship program in bibliography, research scholarship, and staff development. The major thrust of the program is to provide fellowships for younger scholars who will prepare bibliographic studies on topics well represented in the Library's collections.

Dr. Anita Guerrini, visiting assistant professor of the history of medicine at the University of Minnesota, was one of our two Mellon Fellows in Bibliography in 1984–85. Her ably annotated bibliography of the Library's imprints of North and South American natural history, together with its valuable introduction, should be of interest to a broad range of seasoned scholars and graduate students. Dr. Guerrini's previous research and writing on eighteenth-century Newtonianism and her bibliographical work at the National Library of Medicine made her especially well qualified to survey this most important aspect of the Library's collections. Mention should also be made of Mark V. Barrow, Jr., of our staff, who skillfully helped bring this volume to publication.

We would like to acknowledge the generous support of The Andrew W. Mellon Foundation that made possible this work and those that will follow.

January 1986

Edward C. Carter II
Librarian
American Philosophical Society

Foreword

The Andrew W. Mellon Foundation, seeking to increase the scholarly usefulness of our collections and enrich the intellectual life of the Library, made a grant in 1982 to the American Philosophical Society establishing a fellowship program in bibliography, research scholarship, and staff development. The major thrust of the program is to provide fellowships for younger scholars who will prepare bibliographic studies on topics well represented in the Library's collections.

Dr. Anita Guerrini, visiting assistant professor of the history of medicine at the University of Minnesota, was one of our two Mellon Fellows in Bibliography in 1984–85. Her ably annotated bibliography of the Library's imprints of North and South American natural history, together with its valuable introduction, should be of interest to a broad range of seasoned scholars and graduate students. Dr. Guerrini's previous research and writing on eighteenth-century Newtonianism and her bibliographical work at the National Library of Medicine made her especially well qualified to survey this most important aspect of the Library's collections. Mention should also be made of Mark V. Barrow, Jr., of our staff, who skillfully helped bring this volume to publication.

We would like to acknowledge the generous support of The Andrew W. Mellon Foundation that made possible this work and those that will follow.

January 1986

Edward C. Carter II
Librarian
American Philosophical Society

Acknowledgments

I wish first to thank Edward C. Carter II, Librarian of the American Philosophical Society, for his support and encouragement of this project, and also the Andrew W. Mellon Foundation for their grant to the Library which supported my year in Philadelphia.

The staff at the APS Library were uniformly helpful and friendly, and I was very sorry to leave them. In particular, Roy Goodman, reference librarian, provided many excellent suggestions and references, and associate librarian Hildegard Stephans guided me to computer semi-literacy with infinite patience and good humor. Frank Margeson, APS photographer, took the cover picture. Outside the APS, Frederick B. Churchill was extremely helpful in the formation of my initial proposal, and I have gained much benefit from the works of Joseph Ewan and from a conversation with him last autumn.

Last, but certainly not least, Michael Osborne shared his vast knowledge of this subject with me and helped in innumerable other ways. I hope I can repay him in kind.

Introduction

In the early modern period, natural history was a "covering science" which included every aspect of nature—plants, animals, rocks and minerals, geography, and climate, as well as human beings and their behavior. This bibliography includes works on all these topics, running in time from the *Cosmographicus Liber* of Peter Apianus (1524, #13) to the 1770 translation of Per Kalm's *Travels in North America* (#157). In choosing a cutoff date of 1770, I have attempted to complement Max Meisel's magisterial *Bibliography of American Natural History* which begins with the year 1769. Meisel chose that date, the year of the reorganization of the American Philosophical Society (founded in 1743), to signify the beginning of organized scientific endeavor in the North American colonies, shortly to become the United States. I have organized this bibliography rather differently than Meisel did his. I have included only printed books, pamphlets, and broadsides; since Meisel surveyed periodical literature for the period before 1769, I have omitted it. In addition, I have included works covering the whole of the New World, north and south. In this earliest period of European incursion into the New World, any territorial divisions we now recognize would be anachronistic. The collections in the Library of the American Philosophical Society focus particularly on North America and the English-speaking colonies; therefore this bibliography is weighted in that direction, but several significant works on Latin America are also included.

The definition of "natural history" has occupied many historians, and my attempts to define it in this introduction are not meant to settle the question once and for all.[1] Rather, in the context of the 290 specific works below cited from the period 1524 to 1770, I intend to provide a point of reference from which historians can view this period and draw their own conclusions. In compiling this bibliography, I have made a distinction, probably artificial, between "natural history" and "natural philosophy," that is, between works mainly of observation and description and works of theory and generalization. The dividing line between these two areas of endeavor

1. See Paul Lawrence Farber, "Research Traditions in Eighteenth Century Natural History," in *Lazzaro Spallanzani e la Biologia del Settecento* . . . (Firenze: Olschki, 1983), 397–403.

was never clear-cut, and as the period in question progressed the line became even fainter. This blurring is especially evident in the works of writers who attempted to classify, rather than merely list. A system of classification reflects a certain theoretical outlook. Yet such works are also catalogues, and as such I have included them in this bibliography. My main justification for the inclusion of a work was its specific references to American phenomena, a criterion which excludes general experimental or theoretical works.

In Baconian terms, I have included works which proceed up to the level of Bacon's "first vintage" but not beyond. Robert Boyle's *General Heads for the Natural History of a Country* (#35), published (posthumously) in 1692, implicitly distinguishes the history from the philosophy of nature. His proposals demand that the traveler to foreign parts observe and record but not draw conclusions. Yet, because these observations followed Boyle's guidelines, they would not be random or naive.

Even the earliest observations of the New World were within a context. John Elliott has commented that the "innocent eye" existed rarely, if at all, among early travelers to the New World: "It was not the innocent, but the selective eye which first viewed America."[2] The motives of early observers were many and complex. Explorers sought to glorify themselves and their sponsors; missionaries drew moral edification from the undreamed-of bounty of nature they found, as well as from the prospect of winning new souls. Their points of view, overwhelmingly Eurocentric, served to lessen the impact of the discovery of America on Europeans. Sixteenth-century cosmographies and travel accounts emphasized, in the words of a contemporary translator, "straunge novelties and mervaillous thinges," and readers evidently found more marvels to the east than to the west, for works on the east far outnumbered those on the New World in this period.[3]

Such Eurocentrism strangely affected natural history, which, in this earliest period, was secondary to other pursuits. Travelers tended to see the Old World reflected in the New. William Cronon has noticed that early immigrants to New England believed it to be much more like old England than it actually was.[4] Many, even most, observers tended to identify New World plants and animals with those they already knew, even when the resemblance was slight. They repeated European myths about the behavior of certain animals and the effects of certain plants in the new context.[5] Native Americans provided a never-ending source of wonder and moral edification,

2. John H. Elliott, "Renaissance Europe and America: A Blunted Impact?" in Fredi Chiappelli, ed., *First Images of America* (Berkeley: University of California Press, 1976), 16–17.

3. Margaret Hodgen, "Sebastian Munster (1489–1552): A Sixteenth-Century Ethnographer," *Osiris,* 11 (1954): 508 (quoting Richard Eden), 515.

4. William Cronon, *Changes in the Land* (New York: Hill and Wang, 1983), 8–9.

5. James Masterson, "Travellers' Tales of Colonial Natural History," *Journal of American Folklore,* 59 (1946): 51–67, 174–88.

whether they were seen as noble or ignoble savages, or simply lower steps in the chain of being. Wilma George has argued that, although several New World plants and animals were known from maps and travelers' accounts before 1550, only a few, and those the most unusual (the armadillo was a favorite), appeared in the standard encyclopaedic works of the time.[6] Konrad Gesner included only one American animal, the opossum, in his *Historiae Animalium* of 1551 (#122). Edward Topsell, who drew heavily from Gesner, described perhaps half-a-dozen New World animals in his *History of Four-footed Beasts,* published fifty years later (#270). The great botanist Leonhard Fuchs included only three plants from the New World in his *De Historia Stirpium* of 1542 (#114), and he thought one of them, maize, was Asian.

Sixteenth-century cosmographies provide some of our earliest glimpses of European ideas of the natural history of the New World. These works, of which there are several examples in this bibliography, combined maps of the world with characteristically encyclopaedic commentaries which included natural history, civil history, cosmology, and ethnographic description. Maps themselves are important documents for natural history. Wilma George has pointed them out as a source of early animal illustration,[7] and more generally, they graphically illustrate changing ideas of geography. Cartography had become steadily more sophisticated after the rediscovery by the West of Ptolemy's *Geographia* in the early fifteenth century. The intellectual distance between the stylized, schematic *mappa mundi* current in 1400 and the maps of Ortelius and Mercator in the late sixteenth century is vast.

The Italian commentator Peter Martyr (Pietro Martire d'Anghiera) was the first to suspect, in a letter of 1494, that Christopher Columbus had found not the East Indies but a new world. By 1524, the German cosmographer Peter Apianus represented America in his *Cosmographicus Liber* (#13). His map and accompanying commentary were later revised by his student Gemma Frisius, a mathematics professor at Louvain, whose students included Mercator. The *Cosmographia* of Apianus ran through at least thirteen Latin editions in the sixteenth century as well as several vernacular translations. Sebastian Munster's *Cosmographey* (#209) was equally popular. Perhaps the best example of this genre, however, was Abraham Ortelius's beautiful *Theatrum Orbis Terrarum* of 1570 (#214). In the commentary to his map of the New World, Ortelius attempted to reconcile information from several sources, including explorers' accounts and literary sources. He also recounts Plato's tale of Atlantis as a possible description of America.

Apart from cosmographies, the literature on the natural history of the Americas in the early modern period can be divided into two broad catego-

6. Wilma George, "Sources and Background to Discoveries of New Animals in the Sixteenth and Seventeenth Centuries," *History of Science,* 18 (1980): 80.

7. *Ibid., passim;* see also her *Animals and Maps* (1969).

ries: travelers' accounts and catalogues or encyclopaedias. Toward the end of this period, these two types of literature often merge in the works of naturalists such as John Bartram and of scientific expeditions such as La Condamine's to South America. I have already indicated that travel literature has some flaws as a source of accurate observation and description of natural phenomena. Although John Elliott has found more freshness and directness in the accounts of seafaring men than in those of their academic counterparts, he admits that this freshness may be illusory, "overlaid by a layer of preconceptions."[8] Margaret Hodgen claims that the urge to tell a good story predominated over other considerations: "[Travelers] dwelt above all on the hazardous episodes of their voyages, on their shipwrecks, battles with the savages, and near escapes with death. If, on occasion, descriptions of New World topography, flora, fauna, or people crept into their narratives, the strange and bizarre was emphasized at the expense of the prosaic and carefully examined."[9] Self-aggrandizement and pressure from patrons led early explorers to claim that San Francisco Bay was the Atlantic Ocean and the Chesapeake, the Pacific. La Salle drew a map of the course of the Mississippi River in 1682 which indicated that it curved toward the west and the gold of the Spanish Empire. The continued attempts to find a northwest passage during the eighteenth century provide a particularly vivid example of wishful thinking.[10]

Yet Hodgen's harsh judgment is not entirely fair, for surely it is human nature to prefer the strange to the prosaic. The accounts of La Salle's travels by his comrades Tonti (#269) and Joutel (#156) certainly detail many perils, alongside descriptions of the lands and people they encountered; but their lives were indeed perilous. Next to "travel liars" such as Louis Hennepin, who claimed to have canoed to the mouth of the Mississippi and back in two months (in #137), there are the reasoned and detailed accounts of his fellow clerics, Acosta (#6–7) and Charlevoix (#56–60). A few years before Philip Vincent told of his "battell" with the "Pequet Salvages" (#281), William Wood described New England in more peaceful, if perhaps idealized, detail in his *New-England's Prospect* (#287). We can hardly expect all travel literature of this period to be of equal quality, or to conform to our standards of truth. But among the many different texts we can strike a balance of some sort, and measure the meaning of America in this time.

Missionaries and clergymen were responsible for a great deal of travel literature. Indeed Daniel Mornet has observed that by the eighteenth century

8. Elliott, "Renaissance Europe and America," 16.

9. Margaret Hodgen, *Early Anthropology in the Sixteenth and Seventeenth Centuries* (Philadelphia: University of Pennsylvania Press, 1964), 113.

10. See Percy G. Adams, *Travelers and Travel Liars* (Berkeley: University of California Press, 1962), ch. 3–4, pp. 44–79.

natural history was largely a "science ecclésiastique."[11] Although missionaries' descriptions of native peoples were, more often than not, highly colored by their Christian fervor, most clergymen were astute and intelligent observers, as well as zealous collectors of every kind of information. Even as polemical an account as that of Bartolomeo de las Casas (#48–50), who described the Spanish cruelties to native Mexicans in great detail, contains a wealth of information on many other topics. The English clergyman-naturalist, with whom we are familiar in Gilbert White, had his colonial counterparts in John Banister (#236), Griffith Hughes (#148) and William Smith (#258). The clergyman-polymath reached a peak of development in Cotton Mather (#192–193), and the French priests Barrelier (#16) and Plumier (#229–231) may be included among professional men of science.

By the end of the seventeenth century a new emphasis on natural history and on "scientific" observation begins to appear in travel literature. Scientific societies such as the Royal Society in Britain, founded in 1660, and the French Académie Royale des Sciences, founded in 1666, encouraged and promoted the study of nature. John Banister's letters from Virginia in the 1680s were published in the Royal Society's *Philosophical Transactions,* and Boyle's *General Heads* had been preceded by a long series of "Questions for Travellers" in the same publication. As a result, such unlikely individuals as the former pirate William Dampier (#76–78) prided themselves on being men of science, and even voyages of primarily military objective such as Frézier's (#111–113) published accounts of their scientific observations. The dividing line between the amateur and the professional naturalist, never entirely clear, became increasingly fuzzy during the course of the eighteenth century.

In any case, the so-called professional naturalists, with a few exceptions, relied heavily upon travelers' accounts in compiling their encyclopaedic works. One exception was the Spanish physician Francisco Hernández, who spent several years in Mexico in the 1570s compiling a massive manuscript on New World materia medica, only a fraction of which was published (#139). Ironically, his better-known countryman Nicolas Monardes (#204), whose works were cited by many others, never visited the New World. The sixteenth-century encyclopaedists borrowed extensively from each other: Fuchs's identification of maize as "turcicum frumentum" was repeated by Dodoens (#89–91) and Lobel (#185). In such circumstances, erroneous information could survive unquestioned for several generations. Wilma George noted the 1605 illustration of a footless bird of paradise in the *Exoticorum* of L'Ecluse (#175), drawn from a poorly preserved specimen

11. Quoted in William and Mabel Smallwood, *Natural History and the American Mind* (New York: Columbia University Press, 1941), 60.

which had lost its feet in transit.[12] The species survived as Linnaeus' *Paradisea apodea.*

Botany was the most prominent natural historical activity in the New World, and in the early modern period generally, for several reasons. Most plants were easily studied and transported, in contrast to animals. Renaissance naturalistic thought stressed the interconnectedness of the universe, and in this context plants could neatly symbolize nature as a whole. In addition, and perhaps most important for the New World, plants were useful. Botanicals dominated the materia medica, and Europeans welcomed new food plants: the potato, first illustrated in John Gerard's *Herball* in 1597 (#118) quickly became common in the European diet. Spices, dyestuffs and fibers, as well as ornamental plants, were also sought from the New World.[13]

Although botany began to draw away from its medical origins in the late sixteenth century, most botanists were also physicians, and medicinal uses continued to be a compelling motive for the study of American plants. Tobacco, thought to be a cure-all, appears in virtually every herbal from the end of the sixteenth century onward. Guaiacum and sassafras were also highly regarded, the former as a specific for syphilis, another gift from the New World. Most prized, however, was the antimalarial cinchona bark from which quinine is derived. Known as "Peruvian" or "Jesuit's" bark, cinchona, discovered in the 1630s, was soon widely used in Europe against a variety of fevers.

The trickle of American plants which entered Europe in the sixteenth century steadily increased over the next two centuries. Gardens played a prominent role in this increase; indeed, the significance of European botanic and ornamental gardens for the increase of knowledge about New World botany would be difficult to overstate. Gesner's Zurich garden in the mid-sixteenth century was legendary, and his example in planting exotica was followed in Italy and then elsewhere. John Gerard was the first gardener to issue a complete catalogue (#119) of the contents of his London garden, which included several American plants. As historian Keith Thomas has recently shown, during the course of the seventeenth century gardening became an enormously popular and fashionable activity for all classes of society, and popular demand for exotic ornamental plants followed closely behind more scientific motives for plant-collecting in the Americas.[14] Mark Catesby brought trees from the Carolinas and Florida to England in 1726; by 1730, specimens are listed for sale in the catalogue of the London Society of Gardeners (#202). On perhaps a more scientific level, several avid gardeners, including Peter Collinson, John Fothergill, and Lord Petre, financed John

12. George, "Sources and Background," 93–94.

13. See Jonathan Sauer, "Changing Perception and Exploitation of New World Plants in Europe, 1492–1800," in Chiappelli, *First Images of America* (note 2 above), 813–832.

14. Keith Thomas, *Man and the Natural World* (New York: Pantheon, 1983), ch. V.

Bartram's botanizing in exchange for over 100 species of American plants. Hans Sloane, secretary and later president of the Royal Society, encouraged the donation of exotics to the physic gardens in Oxford and Chelsea, and himself contributed several West Indian varieties.

Thomas has also noted the great intellectual and spiritual significance of gardens in the early modern period. The very titles of such works as John Parkinson's *Paradisi in Sole Paradisus Terrestris* (#216, 218) illustrate "the idea that it would be possible to return to a state of pre-lapsarian grace by cultivation of the soil."[15] A further dimension of symbolism connected this notion to the cultivation and civilization of the New World. The garden as a metaphor for civilization—as opposed to the savagery of the wilderness—is prominent in the literature of the seventeenth century, and a professed goal of the early North American colonists was to make of the wilderness a garden. By classifying and bringing order to the wild productions of the New World, European gardens contributed to this end; and like other aspects of natural history, this was also an activity of religious significance, especially to the Puritan colonists. Cotton Mather found no conflict, but rather self-affirmation, in the title of "Christian philosopher," and he subtitled his book of that name, "a collection of the best discoveries in nature, with religious improvements" (#193). Frederick Jackson Turner later found in the frontier the source of American spiritual values, but in this period quite the opposite was true.[16] In a famous essay, Lynn White found the origins of our notions of environmental exploitation in Judeo-Christian thought. Although specific aspects of his argument have been challenged, it is nonetheless true that taming the wilderness had, in this period, strong religious motivations.[17]

Alongside the gardens were the herbaria, zoos and cabinets. Collection of exotic animals in zoos, although never as prominent an activity as gardening in this era, began around 1600. Konrad Gesner's house in Zurich was a prototype of the natural history cabinets of such collectors as Ole Worm (#288), John Tradescant, James Petiver (#224), and, best-known of all, Hans Sloane. Sloane's cabinet, which included the collections of Petiver and others, became the basis of the British Museum. His herbarium in Oxford contained specimens from several American correspondents, as well as several hundred specimens he had collected in the West Indies in the 1680s.

One result of this collecting fervor was a classifying fervor which culminated in Linnaeus and was again primarily botanical. The early attempts

15. *Ibid.*, 236.

16. Frederick Jackson Turner, "The Significance of the Frontier in American History" (1893), in *The Early Writings of Frederick Jackson Turner*, introduction by Fulmer Mood (Madison: University of Wisconsin Press, 1938), 185–229, especially 227–9.

17. Lynn White, "The Historical Roots of Our Ecologic Crisis," *Science,* 155 (1967): 1203–1207. Cf. William Coleman, "Providence, Capitalism, and Environmental Degradation," *Journal of the History of Ideas,* 37 (1976): 27–44.

at classification of Lobel and others reached an early peak in the work of Gaspard Bauhin (#21–22), whose binomial nomenclature, based on natural affinities, endured for much of the seventeenth century. By 1600, the New World and its productions were recognized to be indeed new, and in need of new categories of classification. Just as the closed geocentric universe was being shattered by Copernicus and his followers, so too the notion of the Old World as a full, complete and interconnected universe was shown to be a myth. In consequence, the act of naming and the significance of language took on a different meaning. To the Renaissance mind, the act of naming had mystic significance, and conferred meaning in several interrelated ways. A name fit an object into an established framework of symbols and correspondences through which its true meaning, one not immediately evident upon observation, emerged. So Gesner's *Historiae Animalium* of 1551 is an "opus philosophis, medicis, grammaticis, philologis, poetis, & omnibus rerum linguarumque variarum studiosis, utilissimum simul iucundissimumque futurum."

By the end of the seventeenth century, as Michel Foucault has explained, words had largely lost their double significance of meaning and symbol. Descartes emphasized clear and distinct ideas and the importance of precise definition, on the model of mathematics. Words became, in Foucault's phrase, simply representative signs, flat and without resonance. Names could therefore be manipulated like numbers, and taxonomists such as John Ray (#236–237) sought to emulate the success of the mathematization of the physical sciences by introducing order and precision into the natural world. The result of such classification would be the discovery of general laws, which, for Ray and most of his contemporaries, would demonstrate the handiwork of God in the universe.[18]

When the newness of the New World was recognized, its objects could not be fit into the symbolic pattern of the old. Observation of these objects needed to be non-symbolic, although this was only recognized very slowly. By the late seventeenth and early eighteenth centuries, however, catalogues of American flora and fauna, based on direct observation, reflect a new mode of thought. In works of such people as Hans Sloane (#254–256) and Maria Sibylla Merian (#194), we find a new kind of natural history description, without accoutrements of symbol or use, which lent itself readily to classification.

If the New World changed European ideas of nature, the Europeans also changed that world drastically. Gilbert Chinard has commented that "America was considered . . . as a country in which could be observed, under most favorable conditions, phenomena which belonged to the historical past

18. Michel Foucault, *The Order of Things* (originally published in French, 1966; translated 1971; rpt. New York: Vintage Books, 1973), ch. 2–3; see especially 71–77.

of Europe and could not be reproduced at will."[19] But this world began to vanish as soon as Columbus set foot in the West Indies. Cronon has shown that this world was not entirely virgin, but it had not been systematically exploited by a technologically sophisticated people.[20] At the turn of the seventeenth century, the poet John Donne called his mistress "My America! My new-found-land . . . My Myne of precious stone, My Emperie."[21] The metaphor of America as consort of Europe had a dark side. Carolyn Merchant has emphasized the vocabulary of rape and domination inherent in Baconian language.[22] As natural history was a Baconian enterprise of collecting, listing, and naming, so its governing philosophy was the Baconian quest for the domination of nature by man. Wilderness had no value in itself, and America was ripe for conquest and taming, for stripping of its riches. The natural history of the New World in this period is, therefore, a record of change and conquest, a catalogue of a vanished world.

This bibliography will, I hope, be of use as a research tool for several disciplines, including the history of science and medicine, literary studies, the history of culture and its social relations, anthropology, and Native American studies, as well as to historians of the Americas. The bibliography is arranged alphabetically by author; a given author's works are arranged chronologically. There are separate chronological and geographical indices at the end.

A Note on the Development of the APS Collections

The early history of the Library of the American Philosophical Society has been chronicled by its former associate librarian, Murphy Smith.[23] He notes that the collection was begun, and largely continued, by gifts from members and others, up to the appointment of John Vaughan as the first Librarian of the Society in 1803. Benjamin Franklin was particularly effective during these early years as a conduit for gifts. He left a portion of his own library to the Society in his will. Unfortunately, the remainder was auctioned off in the early years of the nineteenth century. The Society purchased several volumes at that time, making the APS Library the holder of the largest single group of

19. Gilbert Chinard, "The American Philosophical Society and the Early History of Forestry in America," *Proceedings of the American Philosophical Society,* 89 (1945): 452.

20. Cronon, *Changes in the Land, passim.*

21. John Donne, "Elegie: To His Mistris, Going to Bed," ca. 1605.

22. Carolyn Merchant, *The Death of Nature: Women, Ecology, and the Scientific Revolution* (New York: Harper and Row, 1980), 164–172.

23. Murphy D. Smith, *Oak from an Acorn. A History of the American Philosophical Society Library, 1770–1803* (Wilmington: Scholarly Resources, 1976).

books from Franklin's library; but much of the library was dispersed.[24] I have indicated volumes from Franklin's library in the bibliography.

During John Vaughan's long tenure as Librarian (until 1841), he, assisted by members of the Library Committee, particularly Peter S. Du Ponceau and George Ord, began to collect systematically in the areas the Library now recognizes as peculiarly its own, including science, technology, and linguistics. George Ord, who, among the many offices he held in the Society, served as Librarian from 1842–48, was particularly important in the development of the natural history collections. A naturalist in his own right, editor of Wilson and antagonist of Audubon, Ord's mark on these collections is still highly visible.[25]

More recently, in 1972, the Pennsylvania Horticultural Society chose the Library of the American Philosophical Society as a depository for its collection of rare volumes on botany and gardening. The richness of this collection is indicated in the annotations.

24. On Franklin's library, see Edwin Wolf, "A Key to the Identification of Franklin's Books," *Manuscripts,* 8 (1956): 211–214; *idem,* "The Reconstruction of Benjamin Franklin's Library: An Unorthodox Jigsaw Puzzle," *Papers, Bibliographical Society of America,* 56 (1962).

25. See William Lingelbach, "The Library of the American Philosophical Society," *William and Mary Quarterly,* ser. 3, 3 (1946): 48–69.

KEY TO ABBREVIATIONS

Geographic References. When a given work describes a specific geographical area or areas, these are noted in the citation by the abbreviations below. Where more than one region is listed, the order is that of the region's relative importance in the book. The geographical index located at the end of the bibliography allows the user to search for works concerning a given region.

CB	Caribbean
CD	Canada
EC	East coast of North America
LA	French Louisiana
MA	Central America, including Mexico
MC	Middle colonies of North America, including Pennsylvania, New Jersey, Delaware, Maryland
MW	Old Northwest, Ohio Valley
NA	North America
NE	New England
NW	Pacific Northwest
NY	New York State
SA	South America
SE	Southeastern United States
VA	Virginia
WC	West coast of North America

Bibliographic References. See also "Works Consulted in Compiling the Bibliography."

DAB	*Dictionary of American Biography.* 20 vols. (1928–36).
DNB	*Dictionary of National Biography.* 22 vols. (reprinted 1967–68).
DSB	*Dictionary of Scientific Biography.* 16 vols. (1970–80).
Hunt	Rachel McMasters Miller Hunt, *Catalogue of Botanical Books.* Compiled by Jane Quinby. 2 vols. in 3. (Pittsburgh, 1958–61).
Masterson	James Masterson, "Travellers' Tales of Colonial Natural History," *J. Am. Folklore,* 59 (1946): 51–67, 174–88.
NBU	*Nouvelle Biographie Universelle.* 46 vols. (1852–66).
Sabin	Joseph Sabin et al., *Bibliotheca Americana, from its Discovery to the Present Time.* 29 vols. (New York, 1868–1936).
Smallwood	William and Mabel Smallwood, *Natural History and the American Mind* (New York, 1941).

An Annotated Bibliography of Printed Materials in the Library of the American Philosophical Society

1. [Anonymous]. *Beschryvinge van eenige voorname kusten in Oost- en West-Indiën: als Zueriname, Nieuw-Nederland, Florida, van't Eyland Kuba, Brazil, Suratte, Madagascar, Batavia, Peru en Mexico.* . . . Leeuwarden: Meindert Injema, 1716. CB. Sabin 5046.

Bound with Mocquet (#203) and Rochefort (#243). According to Sabin, the section on New Netherland (pp. 50–76) is taken from Edward Melton, *Zee en land reisen door verscheide gewesten des wereld*s, Amsterdam, 1681, Sabin 47472. Melton's description in turn is from Donck and Montanus.

2. ———. *The great importance of Cape Breton, demonstrated and exemplified, by extracts from the best writers, French and English, who have treated of that colony.* . . . London: John Brindley, sold by C. Corbett and M. Cooper, 1746. CD. Sabin 10730.

Not the same as Bollan (#32). Includes a translation of Charlevoix's account with a folding map. Pamphlet.

3. ———. *Historische und geographische Beschreibung von Neu-Schottland.* . . . Frankfurt and Leipzig: Heinrich Ludwig Bronner, 1750. CD. Sabin 56138.

Translation of *A geographical history of Nova Scotia*, London, 1749. In Sabin under Nova Scotia.

4. ———. *A new and complete dictionary of the arts and sciences . . . together with the kingdoms, provinces, cities, towns, and other remarkable places throughout the world.* . . . London: W. Owen, 1754–55. Not in Sabin.

4 vols., illustrated.

5. Achenwall, Gottfried (1719–1772). *Anmerkungen ueber Nordamerika und ueber dasige grosbritannische Colonien. Aus muendlichen Nachrichten des Herrn B. Franklins.* Frankfurt and Leipzig: n.p., 1769. NA. Sabin 108.

Achenwall, a statistician and historian, taught at the University of Göttingen.

6. Acosta, José de (1539–1600). *Naturall and morall historie of the East and West Indies, intreating of the remarkeable things of heaven, of the elements, mettalls, plants and beasts which are proper to that country: together with the manners, ceremonies, lawes, governements and warres of the Indians, translated by E[dward] G[rimestone].* London: Val. Sims for Edward Blount and William Aspley, 1604. MA. Sabin 131.

Originally published in Latin, Salamanca, 1589. Acosta was a Jesuit missionary in Peru from 1569–1588.

7. ———. *Histoire naturelle et morale des Indes, tant orientales qu'occidentales; où il est traicté des choses remarquables. . . .* Paris: Adrien Tiffaine, 1616. MA. Sabin 125 (original translation).

Earlier French edition 1598; originally published in Latin, 1589. Seven French editions to 1621.

8. Acrelius, Israel (1714–1800). *Beskrifning om de Swenska foersamlingars forna och naerwarande tilstaand, uti det saa Kallade Nya Swerige, sedan Nya Nederland, men nu foer tiden Pensylvanien, samt naestliggande orter wid alfwen De La Ware, Waest-Yersey och New-Castle County, uti Norra America. . . .* Stockholm: Harberg and Hesseberg, 1759. MC. Sabin 133.

Translated in *Mem. Hist. Soc. Penna.,* 11(1874): 17–468. Acrelius was a Lutheran pastor in Christiana (now Wilmington, DE) from 1749 to 1756. "His book is the chief literary monument of the Swedes in Delaware" (*DAB*).

9. Alleon-Dulac, Jean-Louis (d. 1768). *Mélanges d'histoire naturelle.* Lyon: Benoît Duplain, 1765. Not in Sabin.

First published 1762. 6 vols., 12 plates. Alleon-Dulac was a village postmaster in the Lyonnais and an amateur naturalist.

10. Alston, Charles (1683–1760). *Tirocinium botanicum edinburgense.* Edinburgh: W. Sands, A. Murray and J. Cochran for G. Hamilton and J. Balfour, 1753. Not in Sabin.

Originally published 1735? Alston was professor of botany and materia medica at the University of Edinburgh. Part III of this work is an index of plants in the Edinburgh botanical garden, which includes many American plants.

11. ———. *Lectures on the materia medica. . . .* London: Edward and Charles Dilly; Edinburgh: A. Kincaid and J. Bell, 1770. Not in Sabin.

2 vols. Prepared by John Hope, Alston's successor at Edinburgh, from Alston's ms. Smallwood has noted Alston's influence on American naturalists who studied in Edinburgh.

12. [Anson, George?] (1697–1762). *A voyage to the South-Seas, and to many other parts of the world. . . .* London: J. Plumb, 1744. SA. Sabin 1632.

An account of Anson's famous voyage around the world in 1740–44; published in numbers. Illustrated with crude woodcuts.

13. Apianus, Petrus (1495–1552). *Cosmographicus liber Petri Apiani mathematici studiose collectus.* [colophon: Landshut: Johann Weyssenburger, for P. Apianus, 1524]. NA. SA. Sabin 1738.

First edition. Ch. 4, "De America." Maps.

14. ———. *Cosmographia, per Gemmam Phrysium, apud Lovanienses medicum ac mathematicum insignem, restituta. . . .* Antwerp: Arnout Berckmann [colophon: printed by Aegidius Coppenius], 1539. NA. SA. Sabin 1744.

"De America," pt. 2, ch. 4. Edited, with additions, by Gemma Frisius; his additions first appear in the second edition of 1529.

15. ———. *Cosmographia Petri Apiani, per Gemmam Frisium apud Lovanienses medicum ac mathematicum insignem....* Paris: Veuve Gautherot, 1553 [colophon dated 1551]. NA. SA. Sabin 1749.

Another edition of #14.

16. Barrelier, Jacques (1606–1673). *Plantae per Galliam, Hispaniam et Italiam observatae....* Paris: Etienne Ganeau, 1714. Not in Sabin. Hunt 432.

Edited by Antoine de Jussieu. 334 plates, including many American plants. Barrelier was a physician and Dominican friar.

17. Bartram, John (1699–1777). *Observations on the inhabitants, climate, soil, rivers, productions, animals, and other matter worthy of notice. Made by Mr. John Bartram, in his travels from Pensilvania to Onondago, Oswego and the Lake Ontario, in Canada. To which is annex'd, a curious account of the cataracts at Niagara, by Mr. Peter Kalm....* London: J. Whiston and B. White, 1751. MC. NY. CD. Sabin 3868.

First edition of the best-known work of America's first native botanist.

18. ———. Preface in Short, *Medicina Britannica,* 1751, #253 below.

This is Bartram's first printed work, done at Franklin's request.

19. ———. *Journal . . . kept upon a journey from St. Augustine up the river St. John's.* London: W. Nicoll and T. Jefferies, 1769.

Appended to Stork, *Description of East-Florida,* 1769, #265 below.

20. Batman, [Stephan] (d. 1584), ed. *Batman uppon Bartholome, his booke De proprietatibus rerum, newly corrected, enlarged, and amended....* London: Thomas East, 1582. Not in Sabin.

Batman's edition of Bartholomeus Anglicus. Includes a description of "Amazones" of America (pp. 214–215) from André Thévet's *Singularitates Galliae Antarcticae sive Americae,* 1557. From the collection of the Pennsylvania Horticultural Society.

21. Bauhin, Gaspard (1560–1624). *Theatri botanici; sive, historiae plantarum . . . liber primus.* Basel: Johann König, 1658. Not in Sabin.

Edited by Jean Gaspard Bauhin. From the collection of the Pennsylvania Horticultural Society.

22. ———. ⌊*Pinax*⌋ *theatri botanici sive index in Theophrasti Diocoridis Plinii et botanicorum qui a seculo scripserunt opera. . . .* Basel: Johann König, 1671. Not in Sabin. Hunt 318–319.

2 vols. Second edition, first published 1623. Bound with *Prodromus,* second edition (first published 1620). Arranged by natural affinities. From the collection of the Pennsylvania Horticultural Society; interleaved with notes by William Forsyth.

23. [Bellegarde, Jean Baptiste Morvan de] (1648–1734). *A general history of all voyages and travels throughout the Old and New World. . . . By M. du Perier of the Royal Academy.* London: Edmund Curll and Egbert Sanger, 1708. Sabin 21331.

Originally published in French, 1707. In Sabin under du Perier.

24. Bellin, Jacques-Nicolas (1703–1772), *Description géographique des isles Antilles possédées par les Anglais . . . pour le service des vaisseaux du Roy.* Paris: Didot, 1758. CB. Sabin 4553.

Bellin is described as "ingénieur de la Marine . . . il fut chargé de dresser . . . les cartes de toutes les côtes des mers connues" (*NBU*).

25. [Bernard, Jean-Frédéric] (d. 1752). *Recueil de voiages au nord contenant divers mémoires très-utiles au commerce et à la navigation.* Vol. 5. Amsterdam: Jean-Frédéric Bernard, 1720. Sabin 4936.

Volume 5 of 12, published 1715–38. A collection of travel accounts, edited by Bernard. Items entered separately in this bibliography.

26. Bertius, Petrus (1565–1629). *Description d'Amérique, qui est le nouveau monde, tirée des tableaux géographiques.*

In Herrera (#140), *Description des Indes occidentales,* 1622, part V, pp. 229–254. Originally published in *P. Bertij tabularum geographicarum contractarum libri septem,* Amsterdam, 1600, Sabin 5014.

27. Betagh, William. *A voyage round the world. Being an account of a remarkable enterprize, begun in the year 1719....* London: T. Combes, J. Lacy, and J. Clarke, 1728. SA. WC. Sabin 5057.

A critique of Commander George Shelvocke's account of his 1719–1722 circumnavigation, called by Sabin "the last of the buccaneering expeditions." Includes a description of the California coast.

28. [Beverley, Robert] (ca. 1673–1722), *Histoire de la Virginie ... traduite de l'anglois.* Paris: Pierre Ribou, 1707. Imprimé à Orléans. VA. Sabin 5115.

A French translation of Beverley's *History of Virginia,* London, 1705. Illustrated.

29. [———]. *Histoire de la Virginie....* Amsterdam: Claude Jordan, 1712. VA. Sabin 5117.

Reprint of #28.

30. Blackwell, Elizabeth (ca. 1700–1758). *A curious herbal....* London: John Nourse, 1739 [i.e. 1737–39]. Not in Sabin. Hunt 510.

2 vols. Includes American plants in the Chelsea botanic garden. 500 colored plates, published in parts. Hunt calls it "a noble monument to a pioneering lady botanist"; Blackwell drew it to rescue her husband from debtors' prison. From the collection of the Pennsylvania Horticultural Society.

31. Blount, Thomas Pope (1649–1693). *A natural history: containing many not common observations: extracted out of the best modern writers.* London: R. Bentley, 1693. Sabin 5999 (original edition).

First published 1692. A compilation from various sources.

32. [Bollan, William] (d. 1776). *The importance and advantage of Cape Breton, truly stated, and impartially considered. With proper maps.* London: John and Paul Knapton, 1746. CD. Sabin 6215.

Descriptive portion from Charlevoix. Cf. *Great Importance of Cape Breton,* (anon.), #2 above.

33. Boturini Benaduci, Lorenzo (1702–1751). *Idea de una nueva historia general de la America septentrional. . . .* Madrid: Juan de Zuñiga, 1746. MA. Sabin 6834.

Bound with #34.

34. ———. *Catalogo del nuevo historico indiano. . . .* Madrid: Juan de Zuñiga, 1746. MA. Sabin 6833.

Bound with #33. A catalogue of the author's cabinet of Central American Indian artifacts, collected during a sojourn of several years in Mexico.

35. Boyle, Robert (1627–1691). *General heads for the natural history of a country, great or small; drawn out for the use of travellers and navigators. . . .* London: John Taylor and S. Holford, 1692. Sabin 7139.

See Introduction, above. Ch. 8: Virginia.

36. Bradley, Richard (1688–1732). *A general treatise of husbandry and gardening. . . .* London: T. Woodward and J. Peele, 1724. Not in Sabin. Hunt 459.

3 vols., illustrated. Issued in parts, 1723–4. Bradley was named professor of botany at Cambridge in 1724, where his ignorance of Latin and Greek created a scandal. From the collection of the Pennsylvania Horticultural Society.

37. ———. *A general treatise of husbandry and gardening. . . .* London: T. Woodward and J. Peele, 1726. Not in Sabin. Hunt 459.

Second edition of #36. 2 vols. From the collection of the Pennsylvania Horticultural Society.

38. ———. *Dictionarium botanicum: or, a botanical dictionary for the use of the curious in husbandry and gardening. . . .* London: T. Woodward and J. Peele, 1728. Not in Sabin.

2 vols. Includes detailed description of several American plants.

39. ———. *A General treatise of agriculture, both philosophical and practical. . . .* London: W. Johnston et al., 1757. Not in Sabin.

Bradley's letters and short works, including notes on the acclimatization of exotics. Illustrated. From the collection of the Pennsylvania Horticultural Society.

40. Breyne, Jakob (1637–1697). *Exoticarum aliarumque minus cognitarum plantarum centuria prima. . . .* Danzig [Gdansk]: David-Fridericus Rhetius, 1678. Not in Sabin. Hunt 352.

Includes Willem ten Rhijne, "Excerpta ex observationibus suis japonicis physicis et ceteris de fructice thee," pp. vii–xxv at end. Illustrated.

41. Brickell, John (1710?–1745). *The natural history of North-Carolina. With an account of the trade, manners, and customs of the Christian and Indian inhabitants. Illustrated. . . .* Dublin: James Carson, 1737. SE. Sabin 7800.

First edition. Largely plagiarized from Lawson's *History of Carolina,* 1709, #172 below.

42. Brisson, [Mathurin-Jacques] (1723–1806). *Regnum animale in classes IX distributum . . . /Le règne animal divisé en IX classes. . . .* Paris: Cl. Jean-Baptiste Bauche, 1756. Not in Sabin.

Text in French and Latin. Illustrated. Dedicated to Réaumur; Brisson was a demonstrator at Réaumur's cabinet.

43. Bry, Theodor de (1528–1598). *Zehender Theil der Orientalischen Indien begreiffendt eine kurtze Beschreibung der neuwen Schiffart gegen nordt osten, voer die Amerische Inseln in* Chinam *und* Iapponiam, *von einem Engellaender*

Henrich Hudson newlich erfunden. . . . Frankfurt: Matthias Becker, 1613. See discussion in Sabin.

Part 10 of *Small voyages,* edited by Johann-Theodor de Bry (1561–1620). 3 plates, 3 maps.

44. [Burke, Edmund] (1729–1797). *An account of the European settlements in America. In six parts.* London: R. and J. Dodsley, 1758. NA. SA. Sabin 9282.

Second edition, "with improvements." First published 1757. 2 vols. Written by Burke with assistance from his brother, Richard, and William Bourke.

45. [———]. *Histoire des colonies européennes dans l'Amérique, en six parties. Traduit . . . par M. E. . . .* Paris: Merlin, 1767. NA. SA. Sabin 9284.

Translation by Marc-Antoine Eidous of #43 above. 2 vols., maps.

46. [Butel-Dumont, Georges-Marie] (1725–1800?). *Description abrégée des possessions angloises et françoises du continent septentrional de l'Amérique telles qu'elles ont été l'année 1755. . . .* N.p., 1759. NA. Not in Sabin.

Originally published in 1755. Butel-Dumont also wrote *Mémoires historiques sur la Louisiane,* 1753.

47. Campanius Holm, Tomas (d. 1702). *Kort beskrifning om provincien Nya Swerige uit America, som nu foertjden af the Engelske kallas Pensylvania. . . .* Stockholm: Wankijfs, 1702. MC. Sabin 10202.

Translated by Peter DuPonceau in *Mem. Penna. Hist. Soc.,* 3 (1834): 13–166. Tomas Campanius was a grandson of John Campanius (1601–83), a Swedish Lutheran missionary. The latter's account of his voyages from Sweden to Delaware is in his grandson's work.

48. Casas, Bartolomeo de las (1474–1566). *Istoria, ò brevissima relatione della distruttione dell'Indie Occidentali. . . .* Venice: Marco Ginammi, 1630. MA. Sabin 11243.

First published as *Brevissima relación de la destruyción de las Indias,* Seville, 1552. Translated by Giacomo Castellani. Double columns, Italian and Spanish. Bound with #49 below.

49. ———. *Il supplice schiavo Indiano. . . .* Venice: Marco Ginammi, 1636. MA. Sabin 12246.

Bound with #48.

50. ———. *La découverte des Indes Occidentales, par les Espagnols.* Paris: André Pralard, 1697. MA. Sabin 11273.

A translation of four of las Casas' tracts by the Abbé de Bellegarde, "softened in some of the cruel parts, which might have given pain to delicate persons" (Sabin). Cf. #23 above.

51. Catesby, Mark (1683–1749). *The natural history of Carolina, Florida and the Bahama Islands. . . .* London: C. Marsh, T. Wilcox, and B. Stichall, 1754. SE. CB. Sabin 11508. Cf. Hunt 486.

Edited by George Edwards. 2 vols., color plates. French and English. First published 1730–43. "The most famous colorplate book of American plants and animal life" (Hunt). From the collection of the Pennsylvania Horticultural Society.

52. ———. *Hortus Europae Americanus: or, a collection of 85 curious trees and shrubs, the produce of North America. . . .* London: J. Millan, 1767. NA. Sabin 11507.

First edition; according to Sabin, this is the same as the author's *Hortus Britanno-Americanus,* 1763, with a new title page. See Hunt 578. 63 figures on 17 plates, derived from Catesby's *History of Carolina.*

53. Chabert, Joseph Bernard, marquis de (1724–1805). *Voyage fait . . . en 1750 et 1751, dans l'Amérique septentrionale pour rectifier les cartes des côtes de l'Acadie, de l'isle Royale et de l'Isle de Terre-neuve; et pour en fixer les principaux points par les observations astronomiques.* Paris: Imprimerie Royale, 1753. CD. Sabin 11723.

First edition. From the library of Benjamin Franklin.

54. Champlain, Samuel de (1567–1635). *Les voyages de la nouvelle France occidentale, dicte Canada . . . depuis l'an 1603. Jusques en l'an 1629. . . .* Paris: Claude Collet, 1632. CD. Sabin 11839.

"The best edition" (Sabin). Earlier editions 1613, 1619; includes map of the St. Lawrence and its sources.

55. [Chanvalon, Jean-Baptiste Thibault de] (ca. 1725–1786). *Voyage à la Martinique, contenant diverses observations . . . faites en 1751 et dans les années suivantes. Lu a l'Académie Royale des Sciences de Paris en 1761.* Paris: Cl. J.-B. Bauche, 1763. CB. Sabin 11936.

Chanvalon performed a statistical survey of Martinique in 1751–57. This work includes observations on meteorology, topography, and "moeurs."

56. Charlevoix, Pierre-François Xavier de (1682–1761). *Histoire de l'isle espagnole ou de S. Domingue. . . .* Amsterdam: François l'Honoré, 1733. CB. Sabin 12128.

From ms. of Jean-Baptiste Le Pers (1675–1735). 4 vols., maps. A Jesuit, Charlevoix was one of the most prolific and judicious writers on the American colonies in the eighteen century.

57. ———. *Histoire et description générale de la nouvelle France, avec le Journal historique d'un voyage fait par ordre du roi dans l'Amérique septentrionnale.* Paris: Rollin fils, 1744. CD. Sabin 12136.

6 vols., plates, maps. Charlevoix's best-known work. He left for Canada in 1720. Sabin has various publishers.

58. ———. *Journal of a voyage to North-America. Undertaken by order of the French king. Containing the geographical description and natural history of that country, particularly Canada. . . .* London: R. and J. Dodsley, 1761. CD. Sabin 12139.

2 vols.; translation of vol. 3 of #57 above. Map.

59. ———. *Letters to the Dutchess of Lesdiguières; giving an account of a voyage to Canada and travels through that vast country, and Louisiana, to the Gulf of Mexico. . . .* London: R. Goadby and R. Baldwin, 1763. CD. LA. Sabin 12140.

Another edition of #58; possibly a new translation.

60. ———. *The history of Paraguay. Containing . . . a full and authentic account of the establishments formed there by the Jesuits. . . .* London: Lockyer Davis, 1769. SA. Sabin 12133.

2 vols.; abridged translation of his *Histoire du Paraguay,* 1756.

61. Churchill, Awnsham (d. 1728) and John, comps. *A collection of voyages and travels. . . .* London: A. and J. Churchill, 1704. Sabin 13015.

4 vols., illustrated, maps. Folio.

62. [Clarke, William] (fl. 1755). *Observations on the late and present conduct of the French, with regard to their encroachments upon the British Colonies in North America. Together with remarks on the importance of these colonies to Great-Britain.* Boston: S. Kneeland, 1755. CD. NA. Sabin 13470.

Dedication signed William Clark.

63. Clayton, John (1657–1725). [*An account of Mr. John Clayton's voyage to and observations on Virginia*].

In Halley, ed., *Miscellanea Curiosa,* 1708, vol. 3, pp. 281–355, #127 below.

64. Clayton, John (1694–1774). *Flora Virginica.* See Gronovius, *Flora Virginica,* #127 and #129 below.

65. [Clerke, Charles] (1741–1779). *A voyage round the world, in His Majesty's Ship* The Dolphin, *commanded by . . . Commodore Byron . . . by an officer on board the said ship.* London: J. Newbery and F. Newbery, 1767. SA. Sabin 9732.

Under Byron in Sabin. Illustrated. Byron's voyage, 1764–66.

66. Cluver, Philipp (1580–1622). *Introductio in universam geographiam tam veterem quam novam.* . . . London: M. Jenour for John Nicholson, 1711. Sabin 13805.

First published in Leyden, 1624. Maps by Johannes Bunonius. America: pp. 414–419; California, p. 420, is described as "arido, sterili atque deserto."

67. Cockburn, John (fl. 1730–35). *A journey over land from the Gulf of Honduras to the great South-Sea. Performed by John Cockburn and five other Englishmen.* . . . London: C. Rivington, 1735. MA. Sabin 14095.

Cockburn's journey: 1730–31, with Thomas Rounce, Richard Banister, John Holland, Thomas Robinson, and John Ballman. Sabin lists eight editions to 1831.

68. Colden, Cadwallader (1688–1776). *The history of the five Indian nations of Canada, which are dependent on the province of New-York in America, and are the barrier between the English and French in that part of the world.* . . . London: Lockyer Davis, J. Wren, and J. Ward, 1755. CD. Sabin 14275.

2 vols., map. Third edition; first published 1727–47. Colden was a colonial governor of New York and wrote scientific treatises.

69. Coles, William (1626–1662). *Adam in Eden: or, nature's paradise.* . . . London: J. Streater for Nathaniel Brooke, 1657. Not in Sabin. Hunt 269.

Although Coles emphasizes English over foreign plants in this herbal, he includes several American plant remedies. From the collection of the Pennsylvania Horticultural Society.

70. Cooke, Edward (fl. 1708–1712). *A voyage to the South Sea, and round the world, perform'd in the Years 1708, 1709, 1710, and 1711.* . . . London: H.M. for B. Lintot et al., 1712. MA. WC. Sabin 16303.

Woodes Rogers' voyage: 1708–11. Maps. Includes the account of Alexander Selkirk which became the basis for *Robinson Crusoe.* Voyages to California are described on pp. 432 ff.

71. Cornut, Jacques Philippe (1606–1651). *Canadensium plantarum, aliarumque nondum editarum historia, cui adiectum est ad calcem Enchiridion botanicum Parisiense.* . . . Paris: Simon Le Moyne, 1635. CD. Sabin 16809. Hunt 227.

First edition of the first Canadian flora. Thirty American plants are first described here, many of them in the Paris botanical garden. Cornut never visited America.

72. Cortes, Hernando (1485–1547). *Historia de la Nueva-España.* Mexico [City]: Joseph Antonio de Hogal, 1770. MA. Sabin 16938.

Second, third and fourth letters of Cortes, edited by Francisco Antonio Lorenzano.

73. Cowell, John (fl. 1700–1730). *The curious and profitable gardener.* . . . London: Weaver Bickerton and Richard Montagu, 1730. Sabin 17237. Hunt 482.

Part II: "Curious directions for cultivating the choicest fruits of the East and West-Indies at a small expence." From the collection of the Pennsylvania Horticultural Society.

74. Coxe, Daniel (1673–1739). *A description of the English province of Carolana.* . . . [London]: Oliver Payne, 1741. SE. Sabin 17281. Hunt 516.

Previous editions 1722, 1727. Map.

75. daCosta, Emanuel Mendes (1717–1791). *A natural history of fossils.* London: L. Davis and C. Reymers, Printers to the Royal Society, 1757. Not in Sabin

Rocks, not fossils. Only vol. 1 was published, "the author not finding or deserving encouragement" (*DNB*).

76. Dampier, William (1652–1715). *A new voyage round the world. Describing particularly, the Isthmus of America.* . . . London: James Knapton, 1699. MA. Sabin 18374.

Fourth edition, corrected; first published 1697. Considered vol. 1 of Dampier's *Voyages*. Maps; this copy has one map wanting.

77. ———. *Voyages and descriptions.* Vol. II. In Three Parts. London: James Knapton, 1700. MA. Sabin 18375.

Second edition; includes index to first two volumes. Part II: "Campeachy" and the Yucatan; Part III: treatise on winds. Maps.

78. ———. *A voyage to New Holland, etc. In the year, 1699....* London: James Knapton, 1703. SA. Sabin 18376.

Volume 3, part 1, of Dampier's *Voyages.* Part II published 1709. Includes description of Bahia and the Brazilian coast. Illustrated, maps.

79. Delisle, Joseph-Nicholas (1688–1768). *Explication de la carte des nouvelles découvertes au nord de la Mer du Sud.* Paris: Desaint et Saillant, 1752. NW. WC. Sabin 35253.

One of the earliest accounts of Bering's voyage; but Delisle's map, based on his belief in the existence of a northwest passage, is very inaccurate. The brother of the famous cartographer Guillaume Delisle, J.-N. Delisle lived in Russia for several years.

80. Derham, William, ed. (1657–1735). *Philosophical letters between the late learned Mr. Ray and several of his ingenious correspondents, natives and foreigners....* London: William and John Innys, 1718. Not in Sabin. Hunt 443.

Includes correspondence between John Ray and Sloane on the West Indies, and a catalogue of Ray's works. See #236–37 below.

81. Desportes, Jean-Baptiste Poupée (1704–1748). *Histoire des maladies de S. Domingue.* Paris: Lejay, 1770. CB. Sabin 64730 (under Pouppé-Desportes; spelled incorrectly on title page).

3 vols. Vol. 3: "Traité ou abrégé des plantes usuelles de Saint Domingue." According to Sabin, this volume is often lacking.

82. [De Ville, Nicolas] (fl. 1719). *Histoire des plantes de l'Europe, et des plus usitées qui viennent d'Asie, d'Afrique, et d'Amérique.... Divisée en deux tomes et*

rangée suivant l'ordre du Pinax de Gaspard Bauhin. Lyon: Nicolas de Ville, 1719. Not in Sabin.

2 vols., illustrated. Attributed to Bauhin in the APS catalog, but the dedication is signed by de Ville.

83. Dezailler d'Argenville, Antoine Joseph (1680–1765). *L'histoire naturelle éclaircie dans deux de ses parties principales, la lithologie et la conchyliologie. . . .* Paris: de Bure l'aîné, 1742. Not in Sabin.

American shells: pp. 169–70.

84. ———. *L'histoire naturelle éclaircie dans une de ses parties principales, l'oryctologie, qui traite des terres, des pierres, des métaux, des minéraux, et autres fossiles . . . Par M.* ***. Paris: de Bure l'aîné, 1755. Not in Sabin.

Illustrated. Repeats the common notion that there was no iron in North America.

85. Dickenson, Jonathan (fl. 1699). *Ongelukkige schipbreuk en yslyke reystogt, van etlyke Engelschen, in den jaare 1696 van Jamaika in West-Indien, na Pensylvania. . . .* Leyden: Pieter van der Aa, [1707]. SE. Sabin 20017.

First published in English, Philadelphia. 1699; translated by W. Sewel. Concerns a shipwreck in Florida. Publication date from Sabin.

86. Dillenius, Johann Jakob (1687–1747). *Historia muscorum: a general history of land and water, etc. Mosses and corals, containing all the known species. . . .* London: J. Millan, 1768. Not in Sabin.

First published 1741; this is another issue of the 85 plates, with abbreviated indices. Dillenius was the first Sherardian professor of botany at Oxford.

87. Dobbs, Arthur (1689–1765). *An account of the countries adjoining to Hudson's Bay in the north-west part of America . . . with an abstract of Middleton's Journal . . . the whole intended to show the great probability of a north-west passage. . . .* London: J. Robinson, 1744. CD. Sabin 20404.

Map by Joseph La France. Dobbs sponsored Christopher Middleton's 1741–2 voyage in search of a northwest passage, then quarreled with Middleton when he failed in his quest.

88. Dodoens, Rembert (1516–1585). *Florum, et coronariarum odoratarumque nonnullarum herbarum historia.* Antwerp: Christophe Plantin, 1569. Not in Sabin. Hunt 104.

Second, revised edition; first published 1568. Woodcuts by P. van der Borscht. From the collection of the Pennsylvania Horticultural Society.

89. ———. *A niewe herball. . . .* London: Gerard Dewes [colophon: printed by Henry Loe, Antwerp], 1578. Not in Sabin. Hunt 132.

First edition of this translation by Henry Lyte of Dodoens' *Cruydeboeck,* based on the French translation by L'Ecluse. From the collection of the Pennsylvania Horticultural Society.

90. ———. *Stirpium historiae pemptades sex, sive libri XXX.* Antwerp: Christophe Plantin, 1583. Not in Sabin. Hunt 143.

First edition of Dodoens' "last and most comprehensive botanical work" (Hunt). Incorporates material from his earlier works. Illustrated. From the collection of the Pennsylvania Horticultural Society.

91. ———. *Cruydt-boeck. . . .* Antwerp: Plantin-Moretus, 1644. Not in Sabin.

Flemish translation of #90 by Françoys van Ravelingen, including the revisions of L'Ecluse. 6 vols. in 3.

92. Douglass, William (1691–1752). *A summary, historical and political, of the first planting, progressive improvements, and present state of the British settlements in North-America. . . .* Boston: vol. 1: printed and sold by Rogers and Fowle, 1749; vol. 2: D. Fowle, 1753. NA. Sabin 20726.

2 vols. Sabin has 1751 for vol. 2.

93. Doyle, William (fl. 1770). *Some account of the British dominions beyond the Atlantic: containing chiefly what is most interesting and least known with respect to those parts; particularly, the important question about the North West Passage is satisfactorily discussed; with a large map. . . .* London: printed for the author by J. Browne, sold by W. Domville et al., [1770]. NA. CD. Sabin 20799.

Part 1 only.

94. Durante, Castore (1521–1590). *Herbario nuovo . . . ove son figure, che rappresentano la vive piante, che nascono in tutta Europa, et nell'Indie Orientali, et Occidentali. . . .* Venice: I. Giunti, 1636. Not in Sabin. Hunt 229.

First published 1585. Descriptions of plants in the form of poems.

95. Eden, Richard (1521?–1576), ed. *The decades of the newe worlde of West India conteynyng the navigations and conquestes of the Spanyardes. . . .* London: William Powell [colophon: printed by Roberte Toy], 1555. MA. SA. Sabin 1561.

Black letter; a translation by Eden of Peter Martyr's *De orbe novo decades* (first complete edition, 1530), along with extracts from Oviedo, Pigafetta, and others. The first presentation of these works in English.

96. Edwards, George (1694–1773). *Histoire naturelle de divers oiseaux. . . .* London: Royal College of Physicians, 1751. Not in Sabin.

4 vols. in 2. A translation of Edwards' *History of birds,* 1743–51. Includes quadrupeds and fish. Exotics in vols. 3–4.

97. ———. *Gleanings of natural history . . . /Glanures d'histoire naturelle. . . .* London: Royal College of Physicians, 1758–64. Not in Sabin.

3 parts in 2 vols.; French and English in double columns. 362 color copper engravings, drawn and engraved by Edwards. Several American specimens from Fothergill. A continuation of Edwards' *History of Birds.*

98. ———. *Essays upon natural history. . . .* London: J. Robson, 1770. Not in Sabin.

Edwards' essays from the *Philosophical Transactions.*

99. Eliot, Jared (1685–1763). *A continuation of the essay upon field-husbandry, as it is, or may be ordered in New-England. The fourth part.* New York: J. Parker and W. Weyman, 1753. NE. Sabin 22136 (collected edition, 1760).

Part 4 of 6, 1748–59. Discusses varieties of soil.

100. Ellis, Henry (1721–1806). *A voyage to Hudson's-Bay, by the* Dobbs Galley *and* California, *in the years 1746 and 1747, for discovering a North West Passage; with an accurate survey of the coast, and a short natural history of the country. . . .* London: H. Whitridge, 1748. CD. Sabin 22312.

The second voyage sponsored by Arthur Dobbs (see #87). When the northwest passage was not found, Dobbs finally dropped the subject. He later became colonial governor of North Carolina. Map, plates.

101. Ellis, John (1705–1776). *Directions for bringing over seeds and plants, from the East-Indies and other distant countries, in a state of vegetation: together with a catalogue of such foreign plants as are worthy of being encouraged in our American Colonies, for the purposes of medicine, agriculture, and commerce. To which is added, the . . . description of . . . Dionaea muscipula. . . .* London: L. Davis, 1770. Sabin 22319. Hunt 606.

Ellis, an agent in West Florida and Dominica, published several articles in the *Philosophical Transactions.*

102. [Engel, Samuel] (1702–1784). *Mémoires et observations géographiques et critiques sur la situation des pays septentrionaux de l'Asie et de l'Amérique . . . par Mr.* ***. Lausanne: Antoine Chapuis, 1765. WC. CD. Sabin 22571.

Engel was a Swiss geographer and economist. Part 2: Western North America; part 3: possibility of a northwest passage. Maps. A compilation of earlier accounts.

103. Evans, Lewis (1700?–1756). *A map of Pensilvania, New-Jersey, New-York, and the three Delaware counties. . . .* [Philadelphia]: published by Lewis Evans [printed by Benjamin Franklin?], 1749. MC. Not in Sabin.

Engraved by L. Hebert; this copy inscribed by Franklin to John Mitchell. Includes legends about the origins of northeast storms, lightning and electricity, navigation of the Delaware and other bays, etc.

104. ———. *Geographical, historical, political, philosophical, and mechanical essays. The first, containing an analysis of a general map of the middle British colonies in America; and of the country of the confederate Indians: a description of the face of the country; the boundaries of the confederates; and the maritime and inland navigations of the several rivers and lakes contained therein.* Philadelphia: B. Franklin, and D. Hall, 1755. MC. Sabin 23175.

First edition; map. Evans was a well-known colonial surveyor and cartographer.

105. ———. *Geographical, historical, political, philosophical, and mechanical essays. The first....* Philadelphia: B. Franklin, and D. Hall, 1755. MC. Sabin 23175.

Second edition of #104; two copies, one from Franklin's library.

106. Evelyn, John (1620–1706). *Sylvia, or a discourse of forest-trees, and the propagation of timber in His Majesties dominions . . . to which is annexed Pomona . . . also Kalendarium hortense.* London: Jo[hn] Martyn and Ja[mes] Allestry, 1664. NE. Not in Sabin.

First edition. A few comments on New England, expanded in subsequent editions. The first book published by the Royal Society of London. From the collection of the Pennsylvania Horticultural Society.

107. ———. *Sylva, or a discourse of forest-trees....* London: J. Martyn, 1670. NE. Not in Sabin.

Second edition of #106.

108. ———. *Silva, or a discourse of forest-trees....* London: Robert Scott et al., 1706. NE. Not in Sabin.

Fourth edition of #106.

109. Fer, Nicolas de (1646–1720). *Description des isles Antilles,* N.p., n.d. CB. Not in Sabin.

Broadside. Fer was a French geographer and engraver.

110. Feuillée, Louis Econches (1660–1732). *Journal des observations physiques, mathématiques, et botaniques faites . . . sur les côtes orientales de l'Amérique méridionale, et dans les Indes Occidentales . . . 1707 . . . 1712.* Paris: vols. 1–2: Pierre Giffart, 1714; vol. 3: Jean Mariette, 1725. SA. Sabin 24224–5.

3 vols., plates, maps. Volume 3 includes "un autre voiage . . . à la nouvelle Espagne." Contains the first illustrations of ferns from Chile and Peru. Feuillée was a Minimite frair.

111. Frézier, [Amédée François] (1682–1773). *Relation du voyage de la mer du Sud aux côtes du Chily et du Pérou, fait pendant les années 1712, 1713, et 1714. . . .* Paris: Jean-Geoffroy Nyon, Etienne Ganeau, Jacques Quillau, 1716. SA. Sabin 25924.

Frézier was "ingenieur ordinaire" to the king of France.

112. ———. *A voyage to the South-Sea, and along the coasts of Chili and Peru, in the years 1712, 1713, and 1714. . . . Postscript by Dr. Edmund Halley. . . .* London: Jonah Bowyer, 1717. SA. Sabin 25926.

The postscript is damaged in this copy. Translation of #111, with same illustrations.

113. ———. *Relation du voyage . . . avec une réponse à la préface critique du livre intitulé, Journal des observations physiques, mathématiques et botaniques du R.P. Feuillée, contre la Relation du voyage de la mer du Sud. . . .* Paris: Nyon, Didot, Quillau, 1732. SA. Sabin 25925.

A reprint of the 1716 edition (#111) with a response to Feuillée (#110) in a supplement.

114. Fuchs, Leonhard (1501–1566). *De historia stirpium commentarii insignes. . . .* Basel: Isengrim, 1542. Not in Sabin. Hunt 48.

First edition of the first modern botanical glossary. Includes 40 plants never before illustrated, three of these American (maize and two squashes). Woodcuts by Fullmaurer, Meyer and Speckle. From the collection of the Pennsylvania Horticultural Society.

115. ———. *De historia stirpium commentarii insignes. . . .* Leyden: Balthazar Arnoullet, 1549. Not in Sabin. Hunt 61.

First octavo edition of #114, with woodcuts by Clement Boussy.

116. ———. *De historia stirpium. . . .* Leyden: Balthazar Arnoullet, 1551. Not in Sabin.

Another edition of #115. From the collection of the Pennsylvania Horticultural Society.

117. Gage, Tho[mas] (1603?–1656). *A new survey of the West-Indies. . . .* London: M. Clark for J. Nicholson and T. Newborough, 1699. MA. Sabin 26301.

Fourth edition; first published in 1648 as *The English-American.* Gage was a Dominican friar from 1625–37. At a time when Englishmen were banned from the Spanish colonies, he smuggled himself onto a ship in a biscuit barrel in 1625 and remained in Central America for several years. His book provided the first description of much of Central America to non-Spanish-speakers.

118. Gerard, John (1545–1612). *The herball or generall historie of plantes.* London John Norton [colophon: Imprinted at London by Edm. Bollifant, for Bonham and John Norton], 1597. Not in Sabin. Hunt 175.

First edition of this famous work. Much of it is taken from Dodoens, but Gerard included his own comments and arranged it by Lobel's classification. It includes the first illustration of the potato. Two copies, one from the collection of the Pennsylvania Horticultural Society.

119. ———. *Catalogus arborum, fruticum ac plantarum tam indigenarum, quam exoticarum in horto Johannis Gerardi . . . nascentium.* London: Arnold Hatfield, imp. John Norton, 1599. Not in Sabin. Hunt 177.

Second edition of Gerard's catalogue of his London garden. The first edition, 1596, is extremely rare; only one copy is known to exist. From the collection of the Pennsylvania Horticultural Society. Pamphlet.

120. ———. *The herball or generall historie of plants . . . enlarged and amended by Thomas Johnson.* London: Adam Islip, Joice Norton and Richard Whitakers, 1633. Not in Sabin. Hunt 223.

Johnson (1561–1644), apothecary and botanist, corrected many of Gerard's errors and replaced his illustrations with more accurate ones. Two copies, one from the collection of the Pennsylvania Horticultural Society.

121. ———. *The herball; or, generall historie of plantes. . . .* London: Adam Islip, Ioice Norton, Richard Whitakers, 1636. Not in Sabin.

Third edition, the second edited by Johnson. From the collection of the Pennsylvania Horticultural Society.

122. Gesner, Konrad (1516–1565). *Historiae animalium liber primus. De quadrupedibus viviparis. Opus philosophis, medicis, grammaticis, philologis, poetis, et omnibus rerum linguarumque variarum studiosis, utilissimum simul iucundissimumque futurum.* Zurich: Christ. Froschower, 1551. Sabin 27227 (1585–1606 edition).

First edition. One New World animal, the opossum ("simivulpe"). Hand-colored woodcuts.

123. Gmelin, Samuel Gottlieb (1744–1774). *Historia fucorum.* St. Petersburg: Academy of Science, 1768. Not in Sabin.

Seaweed; 33 folding plates. Gmelin, scion of a noted German scientific family, lived in Russia from 1767 until his death.

124. Goedaert, Jan (1617–1668). *Histoire naturelle des insectes, selon leurs differentes mêtamorphoses.* The Hague: Adrien Moetiens, [1700]. Not in Sabin.

3 vols.; vol 2, ch. XV includes "Des insectes de l'Amérique," by Johannes de Mey (1617–1678). Illustrated. Goedaert was one of the earliest entomologists.

125. Gosnold, Bartholomew (d. 1607). *Relations des voyages . . . à la Virginie en 1602 et 1603. Traduite de l'anglois.* VA.

In Bernard, ed., *Receuil de voiages au nord,* 1720, #25 above, pp. 382–408.

126. Great Britain. Parliament. *An act for the preservation of white and other pine-trees growing in Her Majesties colonies of New-Hampshire, the Massachusets-Bay . . . for the masting Her Majesties navy.* London: Thomas Newcomb and Henry Hills, 1711. NE. Not in Sabin.

From the Parliament held 25 November 1710, pp. 387–88.

127. Gronovius, Johannes Fridericus (1686–1762). *Flora Virginica exhibens plantas quas v[ir] c[larissimus] Johannes Clayton in Virginia observavit atque collegit. . . .* Leyden: Cornelius Haak, 1743. VA. Sabin 28923.

2 parts. Clayton's list of Virginia plants, arranged according to the Linnaean system by Gronovius.

128. ———. *Index supellectis lapideae. . . .* Leyden: Cornelius Haak, 1750. Not in Sabin.

Includes specimens of American rocks and minerals. Bound with Baldassari, *Osservationi,* 1750.

129. ———. *Flora Virginica exhibens plantas, quas nobilissimus vir D. D. Johannes Claytonus . . . observavit, collegit et obtulit. . . .* Leyden: n.p., 1762. VA. Sabin 28924. Hunt 571.

Expanded version of 1743 edition (#127), edited by L. T. Gronovius. From the collection of the Pennsylvania Horticultural Society.

130. Gronovius, Laurens Theodorus (1730–1777). *Museum ichthyologicum, sistens piscium indigenorum et quorundam exoticorum. . . .* Leyden: Theodore Haak, 1754. Not in Sabin.

Fascicle 1 only. Fascicle 2, 1756, not in APS. Includes specimens sent by Alexander Garden of Charleston. Gronovius was a son of J. F. Gronovius.

131. [Hales, Stephen] (1677–1761). *The theory and history of earthquakes.* London: J. Newbery et al., [ca. 1755]. Not in Sabin.

Includes descriptions of the Jamaica earthquake of 1692 and the Lima earthquake of 1687. Pamphlet; bound with two accounts of the 1755 Lisbon earthquake. From the library of Benjamin Franklin.

132. [Halley, Edmond] (1656–1742), ed. *Miscellanea curiosa. Containing a collection of some of the principal phaenomena of nature....* London: J. M. for R. Smith, 1708. VA. Sabin 49437 (first edition).

Second edition, first published 1705–7. A collection of items from the *Philosophical Transactions.* 3 vols. Vol. 3 includes John Clayton's letters from Virginia, 1688, #63 above.

133. Hartlib, Samuel (ca. 1600–1662). *Samuel Hartlib his legacie....* London: R. and W. Leybourn for Richard Wodenothe, 1652. Sabin 30702 (first edition, 1651).

Second edition, first published 1651. Includes notes on the domestication of American plants.

134. ———. *Samuel Hartlib his legacy of husbandry....* London: J. M. for Richard Wodenothe, 1655. Not in Sabin. Hunt 264.

Third edition of #133. From the collection of the Pennsylvania Horticultural Society.

135. Hennepin, Louis (1640–1701?). *Description de la Louisiane, nouvellement découverte au sud'ouest de la Nouvelle France, par ordre du roy. Avec la carte du pays: les moeurs et la manière de vivre des sauvages.* Paris: Sebastien Hure, 1683. LA. Sabin 31347.

First edition of the first printed account of Louisiana. Map by Guerard. Henepin, a Jesuit, was in North America from 1678–82.

136. ———. *Descrizione della Luigiana, paese nuovamente scoperto nell'America settentrionale . . . tradotto dal francese....* Bologna: Giacomo Monti, 1686. LA. Sabin 31356.

First translation of #135, by Casimiro Freschot. "It is a little remarkable that the first translation should have been into Italian" (Sabin). Map, wanting.

137. ———. *A new discovery of a vast country in America extending . . . between New France and New Mexico . . . with a continuation; giving an account of the*

attempts of... La Salle upon the mines of St. Barbe. London: M. Bentley et al., 1698. LA. Sabin 31370.

1 of 2 issues, 1698. A translation of *Nouvelle découverte...*, Utrecht, 1697. Plates, maps.

138. ———. *Voyage en un pays plus grand que l'Europe, entre la Mer Glaciale et le Nouveau Mexique.*

In Bernard, ed., *Receuil de voiages au nord,* #25 above. Originally published Utrecht, 1698.

139. Hernández, Francisco (1514–1587). *Rerum medicarum Novae Hispaniae thesaurus, seu plantarum, animalium, mineralium Mexicanorum historia....* Rome: Vitalis Mascardi, for Blasij Deversini and Zanobij Masotti, 1651. MA. Sabin 31516. Hunt 247 (240).

Edited from Hernández' ms. by Nardo Antonio Recchi; includes additions by various authors. See Hunt for collation. First published in this edition, 1628; according to Sabin, this may be a reissue with a new title page. Hernández, a Spanish physician, was in Mexico from 1570–77; the massive ms. upon which this work is based was destroyed in the seventeenth century.

140. Herrera y Tordesillas, Antonio de (1559–1625). *Description des Indes Occidentales qu'on appelle aujourdhuy le Nouveau Monde ... translatée d'espagnol....* Amsterdam: Michel Colin, 1622. MA. SA. Sabin 31543.

Translation of part of Herrera's *Historia general,* 1601–15. Includes Bertius' *Description d'Amérique,* #26 above.

141. ———. *Historia general de los hechos de los castellanos en las islas y tierra firme del mar oceano.* Madrid: Nicolas Rodriguez, 1726–36. MA. SA. Sabin 31546.

4 vols., first published 1601–15. Covers the period 1472–1554 in eight "decades." Includes *Descripcion de las Indias Ocidentales,* 1730. "The best edition" (Sabin).

142. ———. *Historia general de las Indias Ocidentales, ò de los hechos de los castellanos en las islas y tierra firme del mar oceano.* Antwerp: Juan Bautista Verdussen, 1728. MA. SA. Sabin 31545.

Another edition of #141. 4 vols.

143. Heylyn, Peter (1600–1662). *Cosmographie in foure bookes contayning the chorographie and historie of the whole world and all the principall kingdomes, provinces, seas, and isles thereof.* London: Henry Seile, 1652. NA. SA. Sabin 31655.

First edition. America: Book 4, part II, pp. 99–187. 4 vols in 1. Six editions to 1670 listed in Sabin.

144. [Hickeringill, Edmund] (1631–1708). *Jamaica viewed . . . by E. H.* London: John Williams, 1661. CB. Sabin 31693.

Second edition; first edition also 1661. Map by Col. Edward D'Oyley, commander of forces in Jamaica. Hickeringill is described as a "eccentric divine and pamphleteer" in *DNB*.

145. Hill, John (1716?–1775). *A general natural history; or, new and accurate descriptions of the animals, vegetables, and minerals, of the different parts of the world. . . .* London: Thomas Osborne, 1748–52. Not in Sabin.

3 vols., illustrated. Hill, an apothecary, was a prolific author on a variety of topics.

146. ———. *The British herbal: an history of plants and trees, natives of Britain. . . .* London: T. Osborne, J. Shipton, et al., 1756. Not in Sabin. Hunt 557.

Despite the title, American plants are included. Illustrated; arranged according to Ray's classification; at this time, Hill was anti-Linnaean, although he later changed his mind. Two copies, one from the collection of the Pennsylvania Horticultural Society.

147. ———. *Eden: or, a compleat body of gardening. . . .* London: T. Osborne et al., 1757. Not in Sabin. Hunt 559.

Linnaean classification. Supposedly compiled from the papers of Thomas Hale (fl. 1750). Plates. Two copies; one, with hand-colored plates, from the collection of the Pennsylvania Horticultural Society.

148. Hughes, Griffith (1708?–1750?). *The natural history of Barbados. In ten books.* London: printed for the author, 1750. CB. Sabin 33582. Hunt 536.

29 plates from drawings by Ehret. One folio copy, with colored plates, from the collection of the Pennsylvania Horticultural Society. Also a quarto copy with uncolored plates. Hughes, a colonial parson, also published papers in the *Philosophical Transactions.*

149. [Huttich, Johannes] (1490–1544) and [Grynaus, Simon] (1493–1541). *Novus orbis regionum ac insularum veteribus incognitarum una cum tabula cosmographica....* Basel: Jo. Hervagium, 1555. Sabin 34104.

An account of expeditions to the New World, first published in 1532. Map by Sebastian Munster.

150. Jacquin, Nicolas Joseph, baron (1727–1817). *Enumeratio systematica plantarum, quas in insulis Caribaeis vicinaque Americes continente detexit novas....* Leyden: Theodore Haak, 1760. CB. Sabin 35520.

Linnaean classification. Jacquin, intendant of the Royal Gardens at Schönbrunn (Vienna), traveled to the West Indies and Central America in 1754–59 to gather plants for the garden.

151. ———. *Selectarum stirpium Americanarum historia, in qua ad Linnaeanum systema determinatae descriptaeque....* Vienna: Krause [colophon: typis Josephi Kurtzbock], 1763. CB. Sabin 35521. Hunt 579.

184 plates, folio. The plates "form a remarkable series of West Indian and tropical American species" (Hunt). From the collection of the Pennsylvania Horticultural Society.

152. Jonston, John (1603–1675). *Historiae naturalis....* Amsterdam: Johannes Jacobus Fil. Schipper, 1657. Not in Sabin.

2 vols., 249 copper engravings by Matthew Merian. First published in parts, Frankfurt, 1649–53. A compilation from various sources.

153. ———. *Theatrum universale omnifum animalium. . . .* Amsterdam: R. and G. Wetstein. 1718. Not in Sabin.

Another edition of #152, ed. Henrik Ruysch. 2 vols., 270 plates.

154. ———. *Historiae naturalis de arboribus et plantis libri X. . . .* Heilbronn: Franz Joseph Eckebrecht, 1768–69. Not in Sabin.

2 vols. in 1. Engravings by Merian.

155. ———. *Theatrum universale omnium animalium. . . .* Rouen: Veuve Besongne et Jean-Jacques Besongne, 1768. Not in Sabin.

4 vols. as 2. A reprint of #152; 249 plates.

156. Joutel, Henri (1640?–1735). *Journal historique du dernier voyage que feu M. de La Sale fit dans le Golfe de Mexique, pour trouver l'embouchure, et le cours de la rivière de Missicipi . . . rédigé . . . par Monsieur De Michel.* Paris: Estienne Robinot, 1713. LA. Sabin 36760.

First edition. Compiled from Joutel's notes by Jean (?) Michel. Joutel was La Salle's "intendant" from 1684–87.

157. Kalm, Per (1716–1779). *Travels into North America, containing its natural history. . . .* Vol. 1: Warrington: William Eyres, 1770; vols. 2–3: London: printed for the author and sold by T. Lowndes, 1771. NA. Sabin 36989. Hunt 616.

3 vols. First published in Swedish, 1753–61; translated by John Reinhold Forster. Plates, maps. Kalm was in America from 1748–51.

158. [———] and [Mittelburger, Gottlieb] (fl. 1768). *Histoire naturelle et politique de la Pensylvanie, et de l'établissement des Quakers dans cette contrée. Traduite de l'Allemand p[ar] M. D. S.*. Paris: Ganeau, 1768. MC. Sabin 36991. Hunt 599.

Translated from the German by Jacques Rousselot de Surgy. Map.

159. Keith, William (1680–1749). *History of the British plantations in America . . . part 1, containing the history of Virginia; with remarks on the trade and commerce of that colony.* London: S. Richardson, sold by A. Millar, J. Nourse, and J. Gray, 1738. VA. Sabin 37240.

Includes geographical description of Virginia. Keith was colonial governor of Pennsylvania from 1717–26.

160. Kimayer, Thomas (fl. 1705). *Neu-eroeffnetes Raritaeten-Kabinet/ost-west-indianischer und auslaendischer Sachen. . . .* Hamburg: Christian Liebezeit, printed by Philipp Ludwig Stromer, 1705. Sabin 37746.

A catalogue of Kimayer's cabinet. "Quite scarce" (Sabin).

161. Klein, Jakob Theodor (1685–1759). *Historiae avium prodromus. . . .* Lubeck: Jonas Schmidt, 1750. Not in Sabin.

An anti-Linnaean work on birds. Bound with #162.

162. ———. *Quadrupedum dispositio brevisque historia naturalis.* Leipzig: Jonas Schmidt, 1751. Not in Sabin.

Folding plates. Bound with #161.

163. Laet, Joannes de (1593?–1649). *Novus orbis, seu descriptionis Indiae Occidentalis libri XVIII. . . . Novis tabulis geographicis et variis animantium, plantarum, fructuumque. . . .* Leyden: Elzevir, 1633. NA. SA. Sabin 38557.

First published in Dutch, 1625; subsequent editions enlarged.

164. Lafitau, Joseph-François (1670–1746?). *Moeurs des sauvages ameriquains, comparées aux moeurs des premiers temps.* Paris: Saugrain l'âiné, 1724. CD. Sabin 38596.

2 vols., 41 plates. First published 1723 without plates. Lafitau was a Jesuit missionary in Canada.

165. LaFrance, Joseph (1707–1749?). *A new map of part of North America from the latitude of 40 to 68 degrees....* N.p., [ca. 1744]. NA. Not in Sabin.

Duplicate of map in Dobbs, *Account,* #87 above.

166. Lahontan, Louis Armand de Lom d'Arce, baron (1666–1715). *New voyages to North-America ... to which is added, a dictionary of the Algonkine language....* London: H. Bonwicke et al., 1703. NA. Sabin 38644.

2 vols.; engravings by H. Moll. A translation of the French original of the same year. Maps. Lahontan was in New France from 1683–93; he claimed to have found a fictitious "rivière Longue" north of the Wisconsin River, between the Mississippi and Lake Michigan.

167. ———. *Nouveaux voyages ... dans l'Amérique septentrionale....* The Hague: Frères l'Honoré, 1704. NA. Sabin 38639.

2 vols. in 1. Maps, plates. First published 1703; several subsequent editions.

168. Langham, William (fl. 16th–17th c.). *The garden of health. Containing the sundry rare and hidden vertues and properties of all kindes of simples and plants....* London: Thomas Harper, 1633. Not in Sabin. Hunt 224.

Second edition; first published 1597. From the collection of the Pennsylvania Horticultural Society.

169. Laurence, John (ca. 1668–1732). *A new system of agriculture. Being a compleat body of husbandry and gardening....* London: Tho. Woodward, 1726. NA. Not in Sabin. Hunt 466.

Includes information on American trees from John Evelyn's *Sylva,* #106–8 above. Laurence was a country parson. 2 copies, both from the collection of the Pennsylvania Horticultural Society.

170. Laval, Antoine Jean de (d. 1728). *Voyage de la Louisiane ... en l'année mil sept cent vingt: dans lequel sont traitées diverses matières de physique, astronomie, géographie et marine....* Paris: Jean Mariette, 1728. LA. Sabin 39276.

Maps, tables. Laval was a Jesuit and taught at Toulon.

171. La Ville, Jean Ignace de, abbé (1699?–1774). *Etat présent de la Pensilvanie, où l'on trouve le détail de ce qui s'y est passé depuis la défaite du Général Braddock, jusqu'à la prise d'Oswego, avec une carte particulière de cette colonie.* [Paris?], 1756. MC. Sabin 19370 (under Delaville, Abbé).

An abridged translation of William Smith's *Brief View of the conduct of Pennsylvania,* 1756. La Ville added a "Description abrégée de la Pensilvanie" and "Relation contenant la suite de ce qui s'est passé en Pensilvanie."

172. Lawson, John (d.1711). *The history of Carolina; containing the exact description and natural history of that country: together with the present state thereof. And a journal of a thousand miles, travel'd thro' several nations of Indians. . . .* London: W. Taylor and J. Baker, 1714. SE. Sabin 39452.

First published 1709 as *A New Voyage to Carolina;* this is a reissue of that edition with a different title page. Map, plates. Lawson was a surveyor in the Carolinas; he was killed by Indians in 1711.

173. LeBeau, Charles (1701–1778). *Avantures du Sr. C. Le Beau . . . ou voyage curieux et nouveau, parmi les sauvages de l'Amérique septentrionale. Dans lequel on trouvera une description du Canada, avec une relation très particulière des anciennes coutumes, moeurs et façons de vivre des barbares qui l'habitent et de la manière dont ils se comportent aujourd'hui. . . .* Amsterdam: Herman Uytwerf, 1738. CD. Sabin 39582.

2 vols. Masterson comments that this work is "probably fictitious, though Le Beau is known to have visited Canada."

174. L'Ecluse, Charles de (1526–1609). *Rariorum plantarum historia. . . .* [Antwerp]: Plantin-Moretus, 1601. Not in Sabin. Hunt 180.

Sachs called L'Ecluse the best botanical writer of the sixteenth and seventeenth centuries. As professor of botany at Leyden from 1593–1609, he was responsible for the introduction of many American plants into Europe. In this work, vol. 1 of his collected works, L'Ecluse introduced 100 new species. Illustrated; 2 copies, one from the collection of the Pennsylvania Horticultural Society.

175. ———. *Exoticorum libri decem. . . .* [Antwerp]: Plantin-Raphelengius, 1605. Not in Sabin. Hunt 182.

Vol. 2 of his collected works. Includes animals. 2 copies, one from the collection of the Pennsylvania Horticultural Society.

176. Ligon, Richard (fl. 1657). *A true and exact history of the Island of Barbadoes. . . .* London: Peter Parker and Thomas Guy, 1673. CB. Sabin 41058. Hunt 270 (first edition).

First published 1657; this is a reissue with a new title page. Maps, engravings. Description of sugar industry. Bookplate of William Byrd of Westover.

177. Linnaeus, Carolus (1707–1778). *Musa Cliffortiana florens Hartecampi 1736 prope Harlemum.* Leyden: n.p., 1736. Not in Sabin.

Palms, including American. Linnaeus was superintendant of the botanical garden of the merchant George Clifford at Hartecamp, near Haarlem, during his stay in Leyden from 1735–38.

178. ———. *Hortus Cliffortianus, plantas exhibens quas in hortis tam vivis quam siccis, Hartecampi in Hollandia. . . .* Amsterdam: n.p., 1737 [actually 1738]. Not in Sabin. Hunt 504.

Clifford's garden, classified by Linnaeus. Plates drawn by Ehret and engraved by Wandelaar. From the collection of the Pennsylvania Horticultural Society. See Hunt's comments.

179. ———. *Genera plantarum eorumque characteres naturales, secundum numerum, figuram, situm et proportionem omnium fructificationis partium.* Paris: Michel-Antoine David, 1743. Not in Sabin. Cf. Hunt 582.

First published 1737; this is a reprint of the second edition of 1742. "The starting point of modern systematic botany" (Osler, in Hunt).

180. ———. *Materia medica, liber I, De plantis. . . .* Amsterdam: J. Wetstein, 1749. Not in Sabin.

Several American plants.

181. ———. *Species plantarum, exhibentes plantas rite cognitas ad genera relatas. . . .* Vienna: Johann Thomas von Trattner, 1764. Not in Sabin. Cf. Hunt 548.

2 vols., third edition. First published 1753. The APS also has the famous fourth edition, 1797–1830. 2 copies, one from the collection of the Pennsylvania Horticultural Society.

182. ———. *Mantissan plantarum. Generum editionis VI. Et specierum editionis II* Stockholm: Laurentius Salvius, 1767. Not in Sabin. Hunt 596.

Supplement to *Genera plantarum* and *Species plantarum.*

183. Lister, Martin (1638?–1712). *Historiae sive synopsis methodicae conchyliorum . . . editio altera.* Oxford: Clarendon Press, 1770. Not in Sabin.

Second edition, edited by William Huddleston. Includes Lister's notes for a second edition. The first edition, published 1685–92, consisted only of plates, engraved by Lister's wife and daughter.

184. Little, Otis (1712–1754). *The state of trade in the northern colonies considered; with an account of their produce, and a particular description of Nova Scotia.* London: G. Woodfall, 1748. CD. Sabin 41523.

Pamphlet.

185. Lobel, Matthias de (1538–1616). *Icones stirpium, seu plantarum, tam exoticarum, quam indigenarum. . . .* Antwerp: Plantin-Veuve Moretus, 1591. Not in Sabin. Hunt 138 (first edition).

2 vols. in one. Woodcuts by Plantin. Includes Lobel's table of affinities. First published 1581.

186. ——— and Pena, Pierre (ca. 1535–1605). *Nova stirpium adversaria. . . .* Antwerp: Christophe Plantin, 1576. Not in Sabin. Hunt 127.

Second edition; first published 1571. "A milestone of modern botany" (*DSB*). From the collection of the Pennsylvania Horticultural Society.

187. Logan, James (1674–1751). *Experimenta et meletemata de plantarum generatione. Experiments and considerations on the generation of plants.* London: C. Davis, 1747. Sabin 41795. Hunt 529.

First published in Latin, 1739. This edition in Latin and English, translated by John Fothergill. Pamphlet.

188. Lovell, Robert (ca. 1630–1690). ⌊*Panzoo-oruktologia*⌋. *Sive Panzoologicomineralogia. Or a compleat history, of animals and minerals.* . . . Oxford: Hen. Hall for Jos. Godwin, 1661. Not in Sabin.

Lovell, later a physician in Coventry, published the works listed here while a student at Oxford. Among American animals he includes the "unicorne." Bound with #189.

189. ———. ⌊*Panoruktologia*⌋. *Sive Pammineralogicon.* . . . Oxford: W. Hall for Joseph Godwin, 1661. Not in Sabin.

Minerals. Bound with #188.

190. ———. ⌊*Pambotanologia*⌋. *Sive, enchiridion botanicum. Or, a compleat herball.* . . . Oxford: W. H. for Ric. Davis, 1665. Not in Sabin. Hunt 299.

Second edition; first published 1659. Includes an early listing of the plants in the Oxford Physick Garden. From the collection of the Pennsylvania Horticultural Society.

191. Ludwig, Christian Gottlieb (1709–1773). *Definitiones generum plantarum.* . . . Leipzig: Joh. Friedrich Gleditsch, 1760. Not in Sabin. Cf. Hunt 569.

Second, enlarged edition, edited by Georg Rudolf Boehmer. First published 1737.

192. Mather, Cotton (1663–1728). *The life and death of the renown'd Mr. John Eliot, who was the first preacher of the gospel to the Indians in America. With an account of the wonderful success which the gospel has had amongst the heathen in*

that part of the world: and of the many strange customes of the pagan Indians, in New-England. London: John Dunton, 1691. NE. Sabin 46382.

Second edition; first published in Boston, 1691. John Eliot (1604–1690) was the first to translate the Bible into Indian languages.

193. ———. *The Christian philosopher: a collection of the best discoveries in nature, with religious improvements.* London: Eman. Matthews, 1721. Sabin 46253.

Mather, the author of some 450 works, is described in the *DAB* as "a well-bred amateur."

194. Merian, Maria Sibylla (1647–1717). *Dissertatio de generatione et metamorphosibus insectorum surinamensium . . . /Dissertation sur la génération et les transformations des insectes de Surinam. . . .* The Hague: Pieter Gosse, 1726. CB. Sabin 47960.

Latin and French in double columns. 72 pp., 72 plates, engraved by Mulder and Sluyter. Includes the habitats of the insects. First published 1705 in Latin, reprinted 1719. Maria Sibylla Merian was the daughter of the engraver Matthew Merian. She traveled to Surinam between 1698 and 1701. Both her daughters also became painters.

195. Micheli, Pier Antonio (1679–1737). *Nova plantarum genera iuxta Tournefortii methodum disposita. . . .* Florence: Bernardi Paperinii, 1729. Not in Sabin. Hunt 480.

1900 species, 1400 new (mainly fungi). American species from Sloane and Petiver. 108 plates. From the collection of the Pennsylvania Horticultural Society.

196. Miller, Joseph (fl. 1722–1748). *Botanicum officinale; or a compendious herbal. . . .* London: E. Bell, J. Senex, W. Taylor, and J. Osborn, 1722. Not in Sabin. Hunt 454.

Materia medica. Ray's classification. Dedicated to Sloane.

197. Miller, Philip (1691–1771). *The gardeners dictionary: containing the methods of cultivating and improving the kitchen, fruit and flower garden . . . interspersed with the history of the plants. . . .* London: printed for the author and sold by C. Rivington, 1737. Not in Sabin.

Third edition, first published 1731. Miller has been called "the most important horticultural writer of the eighteenth century" (*DSB*). He was the gardener at the Chelsea Botanic Garden from 1722 to 1770. From the collection of the Pennsylvania Horticultural Society.

198. ———. *The gardeners dictionary. . . .* London: printed for the author and sold by John and James Rivington, 1752. Not in Sabin.

Sixth edition, revised. From the collection of the Pennsylvania Horticultural Society.

199. ———. *The gardeners dictionary. . . .* London: printed for the author and sold by John Rivington et al., 1759. Not in Sabin. Hunt 563.

2 vols., seventh edition, revised. Although this is the first edition to appear after Linnaeus' *Species plantarum,* Miller did not adopt his binomial nomenclature until the next edition, 1768. From the collection of the Pennsylvania Horticultural Society.

200. ———. *Figures of the most beautiful, useful, and uncommon plants described in the Gardeners dictionary, exhibited on three hundred copper plates. . . .* London: printed for the author and sold by John Rivington et al., 1760. Not in Sabin. Hunt 566.

2 vols. Colored plates. Described according to the classifications of Ray, Tournefort and Linnaeus. Two copies, both from the collection of the Pennsylvania Horticultural Society.

201. ———. *Philipp Millers . . . Abbildungen der . . . Pflanzen welche in seinem Gaertner-Lexicon vorkommen . . . aus dem Englischen uebersezt.* Nuernberg: Adam Wolfgang Winterschmidt, 1768–82. Not in Sabin.

2 vols.; colored plates. Translation of #200.

202. [———], comp. *Catalogus plantarum . . . a catalogue of trees, shrubs, plants, and flowers, both exotic and domestic, which are propagated for sale, in the gardens near London . . . by a Society of Gardeners.* London: Society of Gardeners, 1730. Not in Sabin. Hunt 485.

21 color plates; this is one of the earliest flower books with colored plates. Includes many American plants, with details of their export and cultivation. 2 copies, both from the collection of the Pennsylvania Horticultural Society.

203. Mocquet de Meaux, Jean (fl. 17th c.). *De grote nieuw-bereisde wereld: begrypende zes reizen zo na Lybien, d'Eilanden van Kanarien en Barbaryen; Stroom der Amazonen, Karipouzen en Karibanen. . . .* Leeuwarden: Jan Klasen, 1717. SA. CB. Sabin 49792.

Originally published in French, 1616. Bound with Rochefort (#243).

204. Monardes, Nicolas (1493–1588). *Joyful newes out of the New-found Worlde. Wherein are declared, the rare and singuler vertues of divers herbs, trees, plantes, oyles and stones. . . .* London: E. Allde, by the assigned of Bonham Norton, 1596. MA. SA. Sabin 49946. Hunt 173.

Originally published in Spanish, 1569. Translated by John Frampton; translation first published 1577, second edition 1580. This edition is virtually a reprint of the second. Monardes, a physician and botanist, never visited the New World.

205. Montanus, Arnoldus (1625–1683). *De nieuwe en onbekende weereld: of beschryving van America en't Zuid-Land. . . .* Amsterdam: Jacob Meurs, 1671. NA. SA. Sabin 50086.

First edition; illustrated, maps.

206. [———]. *History of America . . . endeavoring a discovery of the unknown parts of the world, especially of Terra Australis Incognita, or the southern continent, with the history of the conquest of Peru and Mexico. . . .* London, 1682. NA. SA. Sabin 50088 (1670 ed.).

Translation of #205 by J. Ogilby, originally published 1670. Illustrated, maps.

207. Morden, Robert (d. 1703). *Geography rectified: or, a description of the world, in all its kingdoms, provinces, countries ... their ancient and present names, inhabitants, situations, histories, customs, governments ... to which is added a compleat geographical index.* ... London: Robert Morden and Thomas Cockerill, 1693. Sabin 50535.

Third edition. Originally published 1680; second edition 1688, fourth edition 1700. Illustrated. Morden was a map and globe maker in London.

208. Müller, [Gerhard Friedrich] (1705–1783). *Voyages from Asia to America, for completing the discoveries of the north west coast of America. To which is prefixed, a summary of the voyages made by the Ru[s]sians on the frozen sea, in search of a north east passage ... translated from the High Dutch of S. Müller.* ... London: T. Jefferys, 1764. NW. WC. Sabin 51285.

Translated by Thomas Jefferys from the third volume of Müller's *Sammlung Russischer Geschichte,* 1758. Second edition of this translation; first published 1761. 4 maps. Jefferys (d. 1771), was a publisher and map engraver who translated several works about the New World. Müller, a German historian, resided in Russia from 1725.

209. Munster, Sebastian (1489–1552). *Cosmographey oder Beschreibung aller Laender, Herrschafften fürnemsten Stetten, Geschichten, Gebreuchen, Hantierungen etc.* ... Basel: Heinrich Petri, 1564. NA. SA. Sabin 51390.

New edition, originally published 1541. A Latin translation appeared in 1550. Illustrated, maps; America: pp. 1424–33. This work, among the best of the sixteenth-century cosmographies, was immensely popular; forty-six editions in six languages appeared before 1650.

210. N. *Voyages aux côtes de Guinée et en Amérique. Par Mr. N***.* Amsterdam: Etienne Roger, 1719. CB. Sabin 51677.

Maps; the plates are mainly of natural history.

211. Neal, Daniel (1678–1743). *The history of New-England containing an impartial account of the civil and ecclesiastical affairs of the country to ... 1700.* London: J. Clark, R. Ford, and R. Cruttenden, 1720. NE. Sabin 52140.

2 vols., map. Ch. 1 includes description of natives; ch. 14 includes natural historical description. Neal was a dissenting minister in London. From the library of Benjamin Franklin.

212. [Oldmixon, John] (1673–1742). *Gross-britannisches America, nach seiner Erfindung, Bevölckerung und allerneuestem Zustand. . . .* Hamburg: Zacharias Hertels, 1710. NA. Sabin 57158.

A translation by Ludwig Friedrich Vischer of *The British Empire in America,* 1708. Sabin suggests this may be a translation of Robert Blome's *English Empire in America.*

213. ———. *The British empire in America, containing the history of the discovery, settlement, progress, and state of the British colonies on the continent and island of America. . . .* London: Brotherton et al., 1741. NA. Sabin 57157.

2 vols. Second edition, first published 1708; this edition differs in content from the first. Maps.

214. [Ortelius, Abraham] (1527–1598). *Theatrum orbis terrarum.* Antwerp: Aegidius Coppenius [printed by Christophe Plantin], 1570. NA. SA. Sabin 57693.

56 maps, colored. First edition; includes *Additamentum,* 1573, with 17 additional maps. Title page wanting. Most engravings by Frans Hoogenberg. America: folio 2. "The first and most rare edition of one of the most celebrated geographical works of the 15th [*sic*] century" (Sabin).

215. Osborne, Thomas (d. 1767), comp. *A collection of voyages and travels . . . compiled from the . . . library of the late Earl of Oxford. . . .* London: Thomas Osborne, 1745. Sabin 57765.

2 vols., both 1745 (Sabin has vol. 2: 1747). Plates, maps. Osborne, a major bookseller in this period, bought the library of Robert Harley, Earl of Oxford, in 1742.

216. Parkinson, John (1567–1650). *Paradisi in sole paradisus terrestris. Or, a garden. . . .* London: Humfrey Lownes and Robert Young, 1629. Not in Sabin. Hunt 215.

109 woodcuts, descriptions of 1000 plants. Parkinson, botanist to Charles I, noted in his preface that many plants had been brought from overseas since Gerard's time; he introduced several new species to England. Hunt calls this book "one of the most beloved of all early English books on gardening." 2 copies, both from the collection of the Pennsylvania Horticultural Society.

217. ———. *Theatrum botanicum: the theater of plants. Or, an herball of a large extent.* . . . London: Tho. Cotes, 1640. Not in Sabin. Hunt 235.

Includes nearly 4000 plants, in comparison to 2850 in Johnson's edition of Gerard. Illustrated. Classified in seventeen "tribes." On the allegorical title page, America is pictured riding on a sheep, holding a bow and arrow. Five copies, all from the collection of the Pennsylvania Horticultural Society (title pages vary slightly).

218. ———. *Paradisi in sole paradisus terrestris. Or, a choise garden.* . . . London: R. N., sold by Richard Thrale, 1656. Not in Sabin. Hunt 267.

Second impression of #216, "corrected and enlarged." From the collection of the Pennsylvania Horticultural Society.

219. Pechey, John (1655–1716). *The compleat herbal of physical plants.* . . . London: R[ebecca] and J[ames] Bonwicke, 1707. Not in Sabin. Hunt 415.

Second edition. Originally published ca. 1694. Part 2: "The exotic or foreign physical plants."

220. Penn, William (1644–1718). *Some proposals for a second settlement in the province of Pennsylvania.* [London]: Andrew Sowle, 1690. MC. Sabin 59735.

Broadside; describes land along the Susquehanna, site of the proposed new settlement.

221. ———. . . . *A general description of the said province [of Pennsylvania], its soil, air, water, seasons and produce, both natural and artificial, and the good increase thereof. With an account of the natives, or aborigines.* London: J. Sowle, 1726. MC. Sabin 59712 (original edition, 1683).

In *A collection of the works of William Penn,* vol. 2, pp. 699–706. Originally published as a pamphlet, 1683.

222. Perrault, Claude (1613–1688). *Memoirs for a natural history of animals. Containing the anatomical descriptions of several creatures dissected by the Royal Academy of Sciences at Paris....* London: Joseph Streater to be sold by T. Basset et al., 1688. Not in Sabin.

Translated by Alexander Pitfeild; illustrated. Originally published in French, 1676. Describes the animals of the Royal Menagerie in Paris who died, includes a few American animals.

223. ———. *Memoirs for a natural history of animals....* London: R. Smith, 1702. Not in Sabin.

Reprint of #222.

224. Petiver, James (1663–1718). *Opera, historiam naturalem spectantia; or, gazophylacium....* London: various publishers, 1695–1713. Sabin 61288.

Issued in 11 parts; plates. This copy was a gift of Sloane to John Bartram, 1744. For collation, see the catalogue of the British Museum (Natural History), vol. 4, p. 1557.

225. [Pichon, Thomas] (1700-1781). *Genuine letters and memoirs relating to the natural, civil and commercial history of... Cape Breton, and Saint John, from the first settlement there, to the taking of Louisburg by the English, in 1758....* London: J. Nourse, 1760. CD. Sabin 62611.

Translation; first published in French, The Hague, 1758. From the library of Benjamin Franklin; includes a list of items in Franklin's hand on the last page.

226. Piso, Willem (1611–1678). *Historia naturalis Brasiliae: de medicina brasiliensi libri quatuor... et G. Marcgravii... historiae rerum naturalium Brasiliae libri octo... cum appendice de Tapuyis, et Chilensibus, J. de Laet....* Leyden: Franciscus Haak; Amsterdam: Ludovicus Elsevir, 1648. SA. Sabin 63028.

Includes section on natural history of Georg Marcgraf de Liebestad (1610–1644) and an appendix by Joannes de Laet (see #163). Piso was physician to the Dutch settlement in Brazil from 1636–44.

227. ———. *De Indiae utriusque re naturali et medica libri quatuordecim....* Amsterdam: Elsevirios, 1658. SA. Sabin 63029.

Enlarged edition of #226. Includes works by Marcgraf and Jacobus Bontius (1592–1631).

228. Plukenet, Leonard (1642–1706). *Opera.* London: T. Davies et al., 1769. Not in Sabin.

4 vols. in 5; 2,740 figures. A reprint of the original edition, 1691–1705. Includes the contents of Plukenet's herbarium.

229. Plumier, Charles (1646–1704). *Description des plantes d'Amêrique. Avec leurs figures.* Paris: Imprimerie Royale [colophon: par les soins de Jean Anisson], 1693. CB. Sabin 63455. Hunt 389.

First edition. Plumier, a Minimite friar and friend of Tournefort, was Royal Botanist for the French islands. He traveled to America three times between 1689 and 1695. 108 plates.

230. ———. *Nova plantarum americanarum genera.* Paris: Jean Boudot [printer to Académie Royale des Sciences], 1703. CB. Sabin 63457. Hunt 407.

Covers the French islands in the West Indies. 106 new genera, each dedicated to a botanist. 40 plates.

231. ———. *Plantarum americanarum fasciculus primus . . . in Insulis Antillis. . . .* Amsterdam: Schouten; Leyden: Gérard Potuliet and Theodor Haak, 1755–60. CB. Sabin 63459 (also as 9343). Hunt 554.

Edited by Johann Burmann, 1706–1779. 262 plates drawn by Plumier, in 10 fascicles. See Sabin for collation.

232. [Poivre, Pierre] (1719–1786). *Voyages d'un philosophe, ou observations sur les moeurs et les arts des peuples de l'Afrique, de l'Asie et de l'Amêrique.* London: n.p.; Lyon: sold by J. de Ville and L. Rosset, 1769. Sabin 63717.

Poivre traveled over most of the world between 1740–57. Sabin has a different publisher. Originally published in 1768.

233. Prade, J. le Royer de (fl. 17th c.). *Tabacks-historia, Insonderheit vom Schnupff-taback.* . . . Schneeberg: Carl Wilhelm Fulden, 1747. Sabin 64875 (1677 ed.). Hunt 308 (1668 ed.).

Originally published in French, 1668. Another French edition, 1677, in Sabin. Pamphlet, incomplete.

234. Prévost d'Exiles, Antoine François (1697–1763), ed. *Histoire générale des voyages, ou nouvelle collection de toutes les relations de voyages par mer et par terre.* . . . Paris: Didot, 1746–89. Sabin 65402.

20 vols. According to Sabin, vols. 1–7 are a translation of vols. 1–4 of Green (Sabin 28539); vols. 8–15 were compiled by Prévost. Index to vols. 1–15 in vol. 16. Maps, plates.

235. Purchas, Samuel (1575?–1626). *His pilgrimage. Or relations of the world and the religions observed in all ages and places discovered, from the creation unto this present . . . a theologicall and geographical historie of Asia, Africa, and America.* . . . London: William Stansby for Henrie Fetherstone, 1614. Sabin 66679.

Second edition, revised. First published 1613. America: Books 8 and 9, pp. 717–918.

236. Ray, John (1627–1705). *Historia plantarum species hactenus editas aliasque insuper multas noviter inventas et descriptas complectens.* . . . London: vols. 1–2: Maria Clark for Henry Faithorne and John Kersey; vol. 3: for S. Smith and B. Walford, 1686–1704. Sabin 68027.

3 vols. Ray's magnum opus. Vol. 2 includes John Banister's catalogue of Virginia plants, and "Compendium historiae plantarum mexicanarum Francisci Hernandez." In vol. 3 there is a list of plants collected by Dampier in Brazil. Illustrated.

237. ———. *L'histoire naturelle, éclaircie dans une de ses parties principales, l'ornithologie.* . . . Paris: Debure père, 1767. Not in Sabin.

A translation by François Salerne of *Ornithologiae,* 1676, based on the researches of Francis Willughby (1635–1672). Plates.

238. Rea, John (d. 1681). *Flora: seu, de florum cultura....* London: T. N. for George Marriott, 1676. Not in Sabin.

Second edition; first published 1665. A book on gardening, this was Rea's only work. From the collection of the Pennsylvania Horticultural Society.

239. ———. *Flora: seu, de florum cultura....* London: G. M., sold by Peter Buck, 1702. Not in Sabin.

Third impression, corrected. From the collection of the Pennsylvania Horticultural Society.

240. Roberts, Lewes (1596–1640). *The merchants mappe of commerce....* London: R. O. for Ralph Mabb, and to be sold by Nicholas Bourne, 1638. Sabin 71906.

First edition. According to Sabin, this is the earliest English work on trade and commerce. Tables, maps. "Of America and the provinces thereof," pp. 53–64.

241. Robson, Joseph. *An account of six years residence in Hudson's Bay, from 1733 to 1736, and 1744 to 1747....* London: J. Payne and J. Bouquet; Edinburgh: Kincaid; Glasgow: Barry; Dublin: J. Smith, 1752. CD. Sabin 72259.

Mainly on trade, but some observations. Appendix I: "Short history of the discovery of Hudson's Bay." Maps. Robson was an engineer and surveyor.

242. [Rochefort, Charles de] (b. 1605). *Histoire naturelle et morale des îles Antilles de l'Amérique....* Rotterdam: Arnout Leers, 1658. CB. Sabin 72314.

First edition. Vocabulary of native tongue by Fr. Raymond Breton. Illustrated. From the collection of the Pennsylvania Horticultural Society.

243. ———. *Natuurlyke en zedelyke historie van d'eylanden, de vooreylanden van Amerika....* Rotterdam: Arnout Leers, 1662. CB. Sabin 72320.

Translation of #242 by H. Dullaert. Bound with Mocquet (#203).

244. [———]. *The history of the Caribby-Islands, viz. Barbados . . . in two books . . . with a Caribbian vocabulary.* London: J. M. for Thomas Dring and John Starkey, 1666. CB. Sabin 72322.

Translation of #242 by John Davies. Illustrated.

245. ———. *Histoire naturelle et morale des îles Antilles de l'Amérique. . . .* Rotterdam: Reinier Leers, 1681. CB. EC. Sabin 72318.

A reissue of the 1665 edition, with a supplement. Written with Louis de Poincy. Supplement: "Récit d'état présent des célèbres colonies de la Virginie, de Marie-Land, de la Caroline, du nouveau duché de York, de Penn-Sylvania, et de la nouvelle Angleterre." Separate title page. 43 pp.

246. Rogers, Robert (1731–1795). *A concise account of North America: containing a description of the several British colonies on that continent, including the islands of Newfoundland, Cape Breton, etc. . . . Also of the interior, or westerly parts of the country. . . .* London: printed for the author, and sold by J. Millan, 1765. CD. MW. Sabin 72723.

Roberts was a major in the French and Indian War, a governor of Michilimackinac, and a royalist during the Revolution.

247. ———. *Journals of Major Robert Rogers. . . .* London: printed for the author, and sold by J. Millan, 1765. NY. Sabin 72795.

Part I, covering 1755–60; part II not published. From Franklin's library.

248. Roggeveen, Arent (d. 1679). *La primera parte del monte de turba ardiente allumbrando con la claridad de su fuego todas las costas firmes, y yslas de toda la India-Occidental. . . .* Amsterdam: Pedro [i.e. Pieter] Goos, 1680. NA. SA. Sabin 72765.

Atlas with descriptive text. First published in Dutch, 1675 ("Het eerste deel van het brandende veen").

249. Salmon, William (1644–1713). *Pharmacopoeia Londinensis: or, the new London dispensatory. . . .* London: J. Dawks for R. Chiswell et al., 1707. Not in Sabin.

Seventh edition of this well-known work. Salmon, an empiric and a prolific writer, traveled in America. From the collection of the Pennsylvania Horticultural Society.

250. ———. *Botanologia: the English herbal; or, history of plants. . . .* London: I. Dawks for H. Rhodes and J. Taylor, 1710 [–11]. Not in Sabin. Hunt 426.

Alphabetically arranged; woodcuts. 2 copies, both from the collection of the Pennsylvania Horticultural Society.

251. Sanson d'Abbeville, Nicolas (1600–1667). *L'Amérique en plusieurs cartes, et en divers traittés de géographie, et d'histoire. Là où sont descripts succinctement, et avec une belle méthode, et facile. Ses empires, ses peuples, ses colonies. . . .* Paris: chez l'autheur, 1657. NA. SA. Sabin 76708.

15 maps, with commentary. Sanson was a mapmaker and royal geographer to Louis XIV. Not the same as #252.

252. ———. *L'Amérique, en plusieurs cartes nouvelles, et exactes, etc. En divers traitez de géographie et d'histoire; là où sont descrits succinctement, et avec une belle méthode, et facile. Ses empires, ses monarchies, ses estats, etc. . . .* Paris: chez l'autheur, [1657]. NA. SA. Sabin 76711.

15 maps and treatises. Not the same as #251.

253. Short, Thomas (1690?–1772). *Medicina Britannica; or a treatise on such physical plants, as are generally to be found in the fields or gardens in Great-Britain . . . with a preface by Mr. John Bartram . . . and his notes throughout the work, shewing the places where many of the described plants are to be found in these parts of America. . . .* Philadelphia: B. Franklin and D. Hall, 1751. Sabin 80573. NA. Hunt 542.

Third edition. See #18 above. According to Sabin, first published in London, 1745.

254. Sloane, Hans (1660–1753). *Catalogus plantarum quae in insula Jamaica sponte proveniunt . . . adjectis aliis quibusdam quae in insulis Maderae, Barbados,*

Nieves, et Sancti Christophori nascuntur. Seu prodromi historiae naturalis Jamaicae. Pars prima. London: D. Brown, 1699. CB. Sabin 82166.

Not illustrated; Sloane's first work on the Caribbean, where he resided from 1687–89 as physician to the governor of Jamaica. He returned to London in 1689 with over 800 species of plants.

255. ———. *A voyage to the islands Madera, Barbados, Nieves, S. Christophers, and Jamaica, with the natural history . . . of the last of those islands. . . .* London: vol. 1: B[enjamin] M[otte] for the author, 1707; vol. 2: for the author, 1725. CB. Sabin 82169. Hunt 417.

2 vols. The earliest representations of the flora and fauna of Jamaica and the neighboring islands. Engravings by Michael van der Gucht and John Savage. Two copies, one from the collection of the Pennsylvania Horticultural Society.

256. [———]. *Histoire de la Jamaïque, traduite de l'anglois. . . .* London: Nourse, 1751. CB. Sabin 82167.

2 vols. Translation of #255.

257. Smith, Samuel (1720–1776). *The history of the colony of Nova-Caesaria, or New-Jersey . . . to . . . 1721; with some particulars since; and a short view of its present state.* Burlington, N.J.: printed and sold by James Parker, also sold by David Hall in Philadelphia, 1765. MC. Sabin 83980.

The author appears to be the same Samuel Smith who wrote *Necessary truth. . .*, 1748.

258. Smith William (fl. 1745). *A natural history of Nevis, and the rest of the English Leeward Charibee Islands in America . . . in eleven letters from the Rev'd. Mr. Smith . . . to the Rev'd. Mr. Mason. . . .* Cambridge: J. Bentham and sold by W. Thurlbourn et al., 1745. CB. Sabin 84563.

Smith was the rector of St. John's, Nevis; Charles Mason was Woodwardian professor at Cambridge.

259. [Smith, William] (1727–1803). *An historical account of the expedition against the Ohio Indians in . . . 1764 . . . including . . . some account of the Indian country . . . by a lover of his country.* Philadelphia: William Bradford, 1765. MW. Sabin 84616.

Bouquet's expedition; attributed to this William Smith in *DAB.* Pamphlet, illustrated, maps. See note in Sabin. William Smith was the first provost of the University of Pennsylvania.

260. Smith, William (1728–1793). *The history of the province of New-York, from the first discovery to the year M.DCC.XXXII. To which is annexed, a description of the country. . . .* London: Thos. Wilcox, 1757. NY. Sabin 84566.

Plates. This William Smith was chief justice of New York from 1763–67, and a loyalist in the Revolution. His correspondence with Colden on this book is in *Coll. N.Y. Hist. Soc.,* 2nd ser., 2 (1848): 193–214.

261. Solis y Ribadeneyra, Antonio de (1610–1686). *The history of the conquest of Mexico by the Spaniards. . . .* London: T. Woodward, J. Hooke, and J. Peele, 1724. MA. Sabin 86487.

First English translation, by Thomas Townsend. Originally published in Spanish, 1684. Folio, plates, maps. Professes to be a continuation of Herrera. Solis was known mainly as a playwright.

262. ———. *Histoire de la conqueste du Mexique, ou de la nouvelle Espagne, par Fernand Cortez. . . .* Paris: Compagnie des libraries, 1730. MA. Sabin 86478.

Fifth edition of the 1691 French translation. According to Sabin, quoting the catalogue of the Bibliothèque nationale, this translation was done by one "S. de Broë, seignor de Citry et de la Guette." In the *Dictionnaire de biographie française* (vol. 7, col. 396), the entry for Bon Broë mentions his great-grandnephews, Bon-François and Augustin, as the last two holders of these titles (both died toward the end of the eighteenth century), and surmises that one or the other is the "S. de Broë" of the catalog of the Bibl. nat. 2 vols., illustrated.

263. Solorzano Pereira, Juan de (1575–1655). *Politica indiana....* Antwerp: Henrico y Cornelio Verdussen, 1703. MA. SA. Sabin 86535.

First published in Latin, 1605. An exhaustive account of native American customs and government. Includes bk. I, ch. 4, "De la naturaleza, excelencias, y cosas raras del Nuevo Orbe."

264. Stith, William (1707–1755). *The history of the first discovery and settlement of Virginia: being an essay towards a general history of this colony.* Williamsburg: Wm. Parks, 1747. VA. Sabin 91860.

Includes much on Indians. Uses John Smith, Beverley, and the Byrd and Randolph papers.

265. [Stork, William] (fl. 1766). *A description of East-Florida, with a journal, kept by John Bartram . . . upon a journey from St. Augustine up the river St. John's, as far as the lakes....* London: W. Nicoll and T. Jefferies. 1769. SE. Sabin 92222.

Third, edition, enlarged. First published 1766; second issue, containing Bartram's journal, also 1766. Maps, plan. Dedication signed William Stork. See #19 above.

266. Stukeley, William (1687–1765). *The philosophy of earthquakes, natural and religious....* London: A. and C. Corbett, 1756. Not in Sabin.

Third edition. Parts I and II first published 1750; part III written after the 1755 Lisbon earthquake. Presentation copy to Benjamin Franklin, 1758; from Franklin's library.

267. [Tabernaemontanus, Jacobus Theodorus] (ca. 1520–1590). *Eicones plantarum, seu stirpium, arborum nempe, fructicum, herbarum, fructuum, lignorum, radicum, omnis generis; tam inquilinorum, quam exoticorum....* Frankfurt-am-Main: Nicolaus Bassaeus, 1590. Not in Sabin.

In his preface the author comments on his many travels, including to America. Illustrated

268. Thomson, Charles (1729–1824). *An enquiry into the causes of the alienation of the Delaware and Shawanese Indians from the British interest . . . together with the remarkable journal of Christian Frederick Post . . . with notes by the editor explaining sundry Indian customs, etc. Written in Pensylvania.* London: J. Wilkie, 1759. MC. Sabin 95562.

Thomson, a merchant and later secretary of the Continental Congress, recorded the treaty of Easton in 1757. Post, a missionary, knew several Indian languages and was involved in treaty negotiations. Folding map. Two copies.

269. Tonti, Henri de (1650–1704). *Relation de la Louisianne; et du Mississippi.* LA.

In Bernard, et., *Receuil de voiages au nord,* 1720, #25 above, pp. 35–198. A reprint of Tonti's *Dernières découvertes dans l'Amérique septentrionale de M. de la Sale. . . ,* Paris, 1697 (Sabin 96172). Tonti was La Salle's closest associate during his explorations of the Mississippi. Tonti himself declared this work to be spurious.

270. Topsell, Edward (d. 1638?). *The history of four-footed beasts and serpents . . . whereunto is now added, the theater of insects . . . by T. Muffet. . . .* London: E. Cotes for G. Sawbridge, T. Williams, and T. Johnson, 1658. Sabin 27228 (under Gesner).

First published 1607–8. Drawn mainly from Gesner. This edition edited by John Rowland, M.D. Includes a few American animals.

271. Ulloa, Antonio de (1716–1795). *Relación histórica del viage à la América meridional. . . .* Madrid: Antonio Marin [and Juan de Zuñiga], 1748. SA. NA. Sabin 36811 (under Juan).

First edition. 2 parts in 4 vols. Co-author: Jorge Juan de Santacilia (1712–1774); but according to Sabin, written by Ulloa. Includes travels in North America. Juan and Ulloa, Spanish naval officers, were members of the French expedition to South America under La Condamine, 1736–43.

272. ———. *Voyage historique de l'Amérique méridionale. . . .* Amsterdam and Leipzig: Arkstee and Merkus, 1752. SA. NA. Sabin 36812.

Translation of #271 by de Mauvillon. 2 vols.; APS has vol. 2 only.

273. [Vallette de Laudun]. *Journal d'un voyage à la Louisiane, fait en 1720. Par M. ***, capitaine de vaisseau du roi.* The Hague and Paris: Musier fils, and Fournier, 1768. LA. Sabin 98393.

Same voyage as Laval's *Voyage de la Louisiane,* 1728, #170 above.

274. Vega, Garcilaso de la (1539–1616). *The royal commentaries of Peru, in two parts.* London: Miles Flesher for Richard Tonson, 1688. SA. Sabin 98760.

An abridged translation, by Sir Paul Rycaut, of *Comentarios Reales,* part 1, 1609, and *Historia general de Peru,* 1616. Garcilaso de la Vega, "El Inca," was the son of a conquistador and an Inca princess.

275 ———. *Histoire des Yncas, rois du Pérou. . . .* Amsterdam: Jacques Desbordes, 1715. SA. Sabin 98752.

2 vols. Translation by Jean Baudouin of *Primera parte de los Comentarios reales,* 1609. This translation first published in 1633. Plates, maps.

276. ———. *La Florida del Inca. Historia del adelantado Hernando de Soto . . . y de otros heroicos caballeros españolas, e indios. . . .* Madrid: Nicolas Rodriguez Franco, Oficina real, 1723. SA. Sabin 98745.

In Andres Gonzalez Barcia, *Ensayo cronológico para la historia general de la Florida* (Sabin 3349). First published in 1605.

277. ———. *Histoire de la conquête de la Floride: ou relation de ce qui s'est passé dans la découverte de ce pais par Ferdinand de Soto. . . .* Leiden: Pieter van der Aa, 1731. SA. Sabin 98748.

Translation of *La Florida del Ynca* by Pierre Richelet, originally published 1670. 2 vols., plates, map.

278. Venegas, Miguel (fl. 1740). *Noticia de la California, y de su conquista temporal, y espiritual hasta el tiempo presente. . . .* Madrid: Manuel Fernandez, 1757. MA. WC. Sabin 98848.

3 vols. Compiled by Andres Marcos Burriel from the manuscripts of Venegas, a Jesuit missionary. "California" refers to Baja California and the extreme southern parts of the present state. Maps.

279. ———. *A natural and civil history of California: containing an accurate description of that country . . . the customs of the inhabitants . . . together with accounts of the several voyages and attempts made for settling California. . . .* London: James Rivington and James Fletcher, 1759. MA. WC. Sabin 98845.

2 vols.; translation of #278. Map, plates.

280. Victoria, Pedro Gobeo de (1560?–1630?). *Argonauticon Americanorum, sive historiae periculorum. . . .* Gdansk [should be Amsterdam?, see Sabin]: Janssonius a Waesberge, 1698. MA. SA. Sabin 99444.

Originally published in Spanish, Seville, 1610. This edition, edited by Johannes Bissel, S. J., first published in Munich in 1647. Map. Victoria was also a Jesuit.

281. [Vincent, Philip] (fl. 1638). *A true relation of the late battell fought in New-England, between the English and the Pequet salvages. . . .* London: Thomas Harper for Nathaniel Butter and John Bellamie, 1638. NE. Sabin 99766.

Second edition; first published 1637. Vincent, a clergyman, first traveled to New England in 1632.

282. Weinmann, Johann Wilhelm (1683–1741). *Phytanthoza iconographia, sive conspectus aliquot millium, tam indigenarum quam exoticarum. . . .* Regensburg: Jerome Lenz, 1737. Not in Sabin. See Hunt 494.

Vol. 1 only. 275 colored plates; this is the first successful botanical book with colored mezzotints. From the collection of the Pennsylvania Horticultural Society.

283. Wheeler, James (fl. 1763). *The botanist's and gardener's new dictionary. . . .* London: W. Strahan et al., 1763. Not in Sabin.

Linnaean classification. From the collection of the Pennsylvania Horticultural Society.

284. Willughby. Francis (1635–1672). *The ornithology of Francis Willughby. . . .* London: A. C. for John Martyn, 1678. Not in Sabin.

Edited by John Ray. An enlarged edition of *Ornithologiae,* 1676. 78 engravings. Cf. #237.

285. Winthrop, John (1714-1779). *A lecture on earthquakes; read in the chapel of Harvard-College in Cambridge, N. E., November 26th 1755. On the occasion of the great earthquake which shook New-England, the week before. . . .* Boston: printed and sold by Edes and Gill, 1755. NE. Sabin 104853.

John Winthrop, scion of the famous New England family, was professor of mathematics at Harvard. Pamphlet.

286. ———. *Relation of a voyage from Boston to Newfoundland, for the observation of the transit of Venus, June 6, 1761.* Boston: printed and sold by Edes and Gill 1761. NE. CD. Sabin 104856.

Pamphlet.

287. Wood, William (1580–1639). *New-England's prospect. Being a true, lively, and experimental description of that part of America, commonly called New-England. . . .* Boston: Thomas and John Fleet, and Green and Russell, 1764. NE. Sabin 105077.

Third edition, originally published 1634. This is a reprint of the second edition, 1639. The earliest topographical description of the Massachusetts colony. Wood lived in New England from 1629–33.

288. Worm, Ole (1588–1654). *Museum Wormianum. Seu historia rerum rariorum. . . .* Leyden: Johannes Elsevier, 1655. Not in Sabin.

A catalogue of the contents of Worm's museum in Copenhagen. Illustrated; published by Worm's son.

289. Wynne, [John Huddlestone] (1743–1788). *A general history of the British Empire in America: containing, an historical, political and commercial view of the English settlements....* London: W. Richardson and L. Urquhart, 1770. NA. Sabin 105682A.

2 vols.; second issue of the first edition, 1769. Map.

290. Zanoni, Giacomo (1615–1682). *Istoria botanica....* Bologna: Gioseffo Longhi, 1675. Not in Sabin. Hunt 342.

First edition, plates. Zanoni was the prefect of the Bologna botanical gardens and professor of botany at the university. From the collection of the Pennsylvania Horticultural Society.

Works Consulted in Compiling the Annotations

Adams, Percy G. *Travelers and Travel Liars.* Berkeley, 1962.

Buckman, Thomas R., ed. *Bibliography and Natural History.* Lawrence, Kansas, 1966.

Butler, June Rainsford. "America—A Hunting Ground for Eighteenth-Century Naturalists." *Papers, Bibliographical Society of America,* 32 (1938): 1–16.

Chiappelli, Fredi, ed. *First Images of America.* Berkeley, 1976.

Chinard, Gilbert. "The American Philosophical Society and the Early History of Forestry in America." *Proceedings of the American Philosophical Society,* 89 (1945): 444–488.

Cronon, William. *Changes in the Land.* New York, 1983.

Crosby, Alfred W. *The Columbian Exchange.* Westport, 1972.

Dictionary of American Biography. 20 vols., 1928–36.

Dictionary of Canadian Biography. Vols. 1–4, 1966–81.

Dictionary of National Biography. 22 vols., reprint 1967–68.

Dictionary of Scientific Biography. 16 vols., 1970–80.

Ewan, Joseph. "Traffick in seeds, shells, and skins from American to European gardens and cabinets, 1578–1860: a selected bibliography." Typescript, 2 pp., 1968.

Freeman, R.B. *British Natural History Books, 1945–1900. A Handlist.* [Folkestone], 1980.

George, Wilma. "Sources and Background to Discoveries of New Animals in the Sixteenth and Seventeenth Centuries." *History of Science,* 18 (1980): 79–104.

Henrey, Blanche. *British Botanical and Horticultural Literature before 1800.* 3 vols. London, 1975.

Hodgen, Margaret T. *Early Anthropology in the Sixteenth and Seventeenth Centuries.* Philadelphia, 1964.

Hunt, Rachel McMasters Miller. *Catalogue of Botanical Books.* Compiled by Jane Quinby. 2 vols. in 3. Pittsburgh, 1958–61.

John Carter Brown Library. *European Americana.* 2 vols. New York, 1980–82.

Masterson, James. "Travellers' Tales of Colonial Natural History." *Journal of American Folklore,* 59 (1946): 51–67, 174–188.

Nash, Roderick. *Wilderness and the American Mind.* 3rd ed., New Haven, 1982.

Neue Deutsche Biographie. 13 vols., continuing, 1953–82.

Nieuw Nederlands biografisch Woorden-Boek. 10 vols., 1911–37.

Nouvelle Biographie Universelle. 46 vols., 1852–66.

Prevost, M. and R. d'Amat, eds. *Dictionnaire de Biographie Française.* Paris, 1956.

Sabin, Joseph, et al. *Bibliotheca Americana. A Dictionary of Books Relating to America, from its Discovery to the Present Time.* 29 vols., New York, 1868–1936.

Smallwood, William and Mabel. *Natural History and the American Mind.* New York, 1941.

Smit, P. and ter Laage, R.J.Ch.V., eds. *Essays in Biohistory Presented . . . to Frans Verdoorn.* Utrecht, 1970.

Stearns, R.P. *Science in the British Colonies of America.* Urbana, 1970.

Stetson, Sarah. "The Traffic in Seeds and Plants from England's Colonies in North America." *Agricultural History,* 23 (1949): 45–56

Geographical Index

This index covers works which contain specific geographical references; therefore, not all all the works in the bibliography are listed. The index is organized by the following geographical areas with the abbreviations used in the bibliography given in parentheses.

Canada (CD)
Caribbean (CB)
Central America, including Mexico (MA)
East coast of North America (EC)
French Louisiana (LA)
Middle colonies of North America, including Pennsylvania, New Jersey, Delaware, Maryland (MC)
North America (NA)
New England (NE)
New York State (NY)
Old Northwest, Ohio Valley (MW)
Pacific Northwest (NW)
South America (SA)
Southeastern North America (SE)
Virginia (VA)
West coast of North America (WC)

Chronological Index